S0-BWR-337

Social History of Nineteenth Century Mathematics

Herbert Mehrtens,
Henk Bos,
Ivo Schneider, editors

1981

Birkhäuser
Boston · Basel · Stuttgart

Editors

Herbert Mehrtens
Technische Universität Berlin
Ernst-Reuter-Platz 7
D-1000 Berlin 10
Federal Republic of Germany

Henk Ros
Math. Instituut
Rijksuniversiteit Utrecht
Budapestlaan 6
The Netherlands

Ivo Schneider
Institut für Geschichte der Naturwissenschaften der Universität München
Deutsches Museum
D-8000 München 26
Federal Republic of Germany

Library of Congress Cataloging in Publication Data
Social history of nineteenth century mathematics.
"Sponsored by the project PAREX"--Pref.
Bibliography: p.
1. Mathematics--History--Congresses.
I. Mehrtens, Herbert. II. Bos, H.J.M.
III. Schneider, Ivo. 1938- . IV. PAREX.
QA21. S74 510'.9 81-10125
ISBN 3-7643-3003-3 AACR2

CIP-Kurztitelaufnahme der Deutschen Bibliothek
Social history of nineteenth century mathematics:
papers from a workshop / Techn. Univ. Berlin, July 5-8, 1979.
Herbert Mehrtens..., ed. -
Boston ; Basel ; Stuttgart : Birkhäuser, 1981.
ISBN 3-7643-3033-3
NE: Mehrtens, Herbert (Hrsg.);Technische Universität
(Berlin, West)

ISBN 3-7643-3033-3
Printed in USA

CONTENTS

APPENDIX

PREFACE

During the last few decades historians of science have shown a growing interest in science as a cultural activity and have regarded science more and more as part of the general developments that have occurred in society. This trend has been less evident among historians of mathematics, who traditionally concentrate primarily on tracing the development of mathematical knowledge itself. To some degree this restriction is connected with the special role of mathematics compared with the other sciences; mathematics typifies the most objective, most coercive type of knowledge, and therefore seems to be least affected by social influences.

Nevertheless, biography, institutional history and history of national developments have long been elements in the historiography of mathematics. This interest in the social aspects of mathematics has widened recently through the study of other themes, such as the relation of mathematics to the development of the educational system. Some scholars have begun to apply the methods of historical sociology of knowledge to mathematics; others have attempted to give a

Marxist analysis of the connection between mathematics and productive forces, and there have been philosophical studies about the communication processes involved in the production of mathematical knowledge. An interest in causal analyses of historical processes has led to the study of other factors influencing the development of mathematics, such as the formation of mathematical schools, the changes in the professional situation of the mathematician and the general cultural milieu of the mathematical scientist.

We feel that these studies, and others which view mathematics as, at least in part, a social activity, may be regarded as "social history of mathematics". A neat definition of such a field cannot and need not be given; we do not intend to announce the start of a new separate subdiscipline. But we want to signal a new, broad and largely interdisciplinary movement in the historiography of mathematics. In this volume we present results of this movement.

The volume consists of papers presented at the third of a series of meetings on social history of mathematics sponsored by the project PAREX.[1)] This third meeting, which took the form of a workshop, was held in West Berlin from July 5th

1) For reports of the meetings see Historia Mathematica 3 (1976), 470-71, 5 (1978), 141-42, 7 (1980), 75-79; Social Studies of Science 8 (1978), 141-42, 10 (1980), 121-125; PAREX informations 2 (1977), 3-4, 3 (1977), 13-14.

to 9th 1979. Twenty-eight participants from eight nations presented and discussed their research, concerning the social history of 19th-century mathematics. Since the texts of the papers were distributed before the conference, there was ample time for discussion. During the meeting there were group discussions on three topics: mathematics in the early nineteenth century, professionalization of mathematics, and methods and research programmes in the social history of mathematics. Reports of these group discussions were presented to the conference and discussed again at a general meeting. As a result of the discussions, which were intense, at times controversial, and very rewarding, most authors decided to revise their papers. Some of the points that were brought up in the discussions have been incorporated in the introductions to the sections. Three of the papers presented at the workshop have been omitted. Two of them have been published elsewhere,[2] and the third [3] is still under study.

We have grouped the papers thematically, as was done during the workshop. Henk Bos wrote the introductions to parts I and III. Part II on the dual topic of education and

2) Grattan-Guiness (1981a, 1981b), Pyenson (1979) (cf. Select Bibliography).

3) MacKenzie, Donald and Mike Barfoot; "Scottish Mathematics - A Sociological Approach".

professionalization, is introduced by Ivo Schneider. Herbert Mehrtens prepared a methodological paper, which served as an introductory exposition for the workshop, and a select bibliography; both are included in the form of an appendix. Herbert Mehrtens organized the editorial division of labour and took upon himself the task of general editor.

We feel that the workshop has been very rewarding. Although the participants had a great variety of disciplinary backgrounds and methodological approaches, the thematic field proved quite homogeneous. The papers present important results of historiographical research as well as an intriguing field for further study.

We are grateful to all those who helped to organize the workshop and who have made it possible to publish the results. We are especially indebted to the staffs of the _Institut für Philosophie, Wissenschaftstheorie, Wissenschafts- und Technikgeschichte_ of the _Technische Universität Berlin_ and of the _Kolpinghaus_, Berlin, where the workshop was held. We also wish to express our gratitude to the _Stiftung Volkswagenwerk_ and to the project _PAREX_, which have given financial assistance, and to the publishing house which has made the publication of this volume possible.

Henk Bos, Herbert Mehrtens, Ivo Schneider.

PART I

ASPECTS OF A FUNDAMENTAL CHANGE

THE EARLY NINETEENTH CENTURY

INTRODUCTION

Henk Bos

The first half of the nineteenth century was a period of great changes in politics, in commerce and industry, in the arts, and in religious, philosophical and scientific thinking. For mathematics as well it was a period of deep change, in views on mathematics as a whole, in ideas about its foundations and the nature of its principal concepts, and in the educational function of the discipline. New institutions for the pursuit and teaching of mathematics and the sciences were created, and older institutions were radically transformed.

The three articles in this section are concerned with these changes, both within mathematics and in society in general. Dirk Struik succinctly formulates the question which underlies the other two articles as well: "How to argue the connection ?".

At the workshop Struik's article also served as an introductory survey. It depicts the change in society and the atmosphere of pioneering, renovation and rebellion which pervaded many aspects of politics and culture. The connection between developments in one field, such as mathematics, and those in other fields and in society in general are not yet well understood, and require further study. But it is Struik's conviction that in a time of such vigorous novelty in so many fields innovations in one field cannot be treated or understood as isolated phenomena.

The other two articles aim at tracing the connections in specific cases. The theme of Luke Hodgkin's paper is the revolution in thinking about the foundations of the calculus, a revolution that was provoked especially by Cauchy's textbooks on the topic. Hodgkin is interested in connections between changes in the practice of mathematics, especially in the education system, which were caused by the political revolution, and changes in the language and the rules of mathematics. In studying this theme he makes use of the concept of "discursive formation" introduced by Michel Foucault in The Archeology of Knowledge. From this methodological starting point Hodgkin discusses two cases. The first is the role of Lacroix's textbooks on the calculus, which, though they date from a period before Cauchy, retained their readability for a long time after Cauchy's textbooks had started the new approach. As another example illustrating the "discursive formation" of the calculus and of mathematics in general in this period,Hodgkin presents and discusses an autobiographical fragment on mathematics by Stendhal.

The article by Niels Jahnke and Michael Otte concerns another deep change in views of the foundations of mathematics which occured during the nineteenth century, namely the process of arithmetization. Jahnke and Otte note that in this period the sciences were directed for the first time towards applications on a broad scale, and that the social basis of science underwent fundamental structural changes in this process. The development of methodological ideas in mathematics, in particular the ideas that played a role in the process of arithmetization, are related to these broader changes within science. Against this background Jahnke and Otte treat the emergence of the concept of relation in mathematics and discuss the arguments of Gauss and Hamilton on the concept of number, in which the role of this concept is central.

The articles of Hodgkin and of Jahnke and Otte are programmatic in the sense that they originate from research projects that are still in progress. They point to further areas in which it would be important to study the connections between changes in society and in the institutional and social bases of science on the one hand and changes in scientific concepts and methods on the other. Such areas were also mentioned in discussions during the workshop, for instance the role of geometry and the influence of textbook literature.

MATHEMATICS IN THE EARLY PART OF THE NINETEENTH CENTURY

Dirk J. Struik

1[1)] Great changes in the social-political structure, and especially revolutions, have a way of influencing the thoughts of men, also in the field of science, including mathematics. The urban revolution of the fourth and third millennium B.C. brought us the Babylonian-Egyptian type of mathematics, the establishment of the Greek polis in the eighth and seventh century B.C. the whole new edifice of Greek science with its new type of mathematics. In modern times revolutions act less slowly. The Revolution in the Netherlands known as the Eighty Years War brought us the mathematics of Stevin and Huygens. The British Revolution, often labeled the Puritan one, carried the Royal Society in its wake, featuring Wallis and Newton. The French Revolution of 1789 equally stimulated a renewal of the mathematical sciences, continued during the Napoleonic period and the Restoration.

There were, of course, two revolutions, the French one, primarily

[1)] This is a somewhat enlarged version of the address presented in Berlin.

political, and the Industrial Revolution, centered in Great Britain. All these revolutions of the sixteenth till the nineteenth century were successive steps in the gradual ascendance to economic and political power of the bourgeoisie, the tiers-état. Its interests and ideals were fostered, stimulated and reflected in the scientific revolution of that period, and mathematics, especially the new mathematics - calculus, analytic geometry , probability - played a fundamental role in this revolution.

The creed of this bourgeoisie, especially of its most powerful and influential section, the haute bourgeoisie, had a tendency toward optimism, leading to a belief in the progress of the human race, progress in knowledge, in power and in social advancement. Beginning with the humanists of Renaissance days - Ulrich von Hutten's "O seculum, o literae, iovat vivere!" - through the seventeenth century's building of great metaphysical systems deeply influenced by mathematics, this belief in progress became the passion of the philosophes beginning with old Fontenelle. Here Newton's mathematical exposition of terrestrial and celestial mechanics showed the ways of God to Man, and led to a belief not only in scientific, but also moral progress. Mathematics, the basis of Cartesian and Newtonian philosophy could not only lead to understanding, but also to economic power, as in the search for the determination of longitude at sea, or the accuracy of artillery. But the mathematical sciences were encouraged above all because of their accumulative character, a visible vindication that progress was possible, and by a bold generalization this tenet was extended to education and morality. As a matter of fact, probability theory showed that mathematics could even be applied to moral issues.

The nineteenth century continued in this belief in progress, but the emphasis was changed. With the growth of other sciences, mathematics lost its rank of pride to be the number one. The belief in progress was now based on the triumphs of the industrial revolution and the science and engineering it promoted - never mind the "dark, satanic mills" of Blake. The decline in this belief among the middle classes came only towards the end of the century with the advent of imperialism with its global conflicts and dark outlook for mankind.

But at the time of the French Revolution belief in progress and in the illuminating example of the mathematical sciences, pures et appliquées, was strong. When the tiers-état seized power and began to reform education for its own needs, it saw in these mathematical sciences a means to this education. And, thanks to the encouragement given to these sciences during the Ancient Regime, the Revolution and the Napoleonic period found a galaxy of leading mathematicians ready to provide both education and paths to new frontiers of science.

One of the most efficient and permanent reforms was, as we know, the opening in 1795 of the Ecole Polytechnique in Paris, created primarily for the education of military engineers for the defense of the young Republic, but also combining other, older, schools as the Ponts et Chaussées, the civil engineering[2]. Military academies were already in existence, as the one in Mézieres near Sedan, where Gaspard Monge had been teaching mathematics and developed his descriptive geometry. The academy at Brienne in Champagne had taught young Bonaparte his love for mathematics.

[2] See e.g. H. Wussing, Die Ecole Polytechnique - eine Errungenschaft der Französischen Revolution. Pädagogik 13 (1958) 646-662.

But the Ecole Polytechnique was a far more ambitious institution, as it developed under the Directoire and Napoleon. With its classroom instruction and general discipline, examination, textbooks developed out of the instruction, the most brilliant scientists as instructors, the school set the example for technical teaching over the Western World and by its stress on the mathematical sciences also deeply influenced university instruction and research. Eventually its influence extended to Prague, Vienna, Stockholm, Zürich, Copenhagen, Karlsruhe, even to the young USA, where West Point, the military academy founded in 1802, was based on the example of the Paris School. It was around this institution, its teachers and its pupils, that Paris maintained for many years its reputation as the mathematical center of the world.

There were other educational reforms in Paris, as the establishment of the Ecole Normale. For many decades teacher training schools in many countries were known as normal schools. The archaic Académie des Sciences was replaced by the Institut - with Napoleon a proud member. But for the development of mathematics we have to look in the first place at the Polytechnique. Here entirely new fields of mathematics were opened.

2 There always will remain persons for whom it is an open question whether, and if so, how, the new flowering of mathematics, so unexpected by the older generation - had not Lagrange written to D'Alembert in 1772 "Ne vous semble-t-il pas que la haute géometrie va un peu à décadence"? - can be related to the political events associated with the great revolution. Founding new schools does not necessarily mean founding new science. It can be maintained that simultaneous events need not be causally connected; and post hoc ergo propter hoc is not a good argument either. Mathematics, it is argued, is an autonomous science, its promotion depends on mathematical genius, and the appearance of genius is ac-

cidental and certainly has nothing to do with politics,commerce or industry. And indeed, the direct demands made by government, mercantile and industrial circles on mathematics during and after the Revolution were not very great, not even in the practice of warfare, especially as compared to the present age. True, many French mathematicians were politically active in some way or another, Carnot was the _organisateur de la victoire_, Monge was a Jacobin and even a _régicide_, Laplace was for a while a minister of marine, Fourier a provincial administrator. In a next generation we find in Galois a militant republican, the opposite of Cauchy, royalist. But the mathematicians of Germany, who contributed so much to the new mathematics were mostly university professors without political ambitions. The role of the mathematical leaders in the new great educational reforms is more pronounced, but this says little of the content of their research. Then, how to argue the connection?

This we can probably do by realizing that the new mathematics was only one aspect of the vigorous pioneering, renovation and rebellion that went on in almost all aspects of intellectual and artistic, literary, religious, moral and scientific thinking of Europe, wherever the armies of the republic and empire had brought the slogans of liberty, equality and fraternity to every nook and corner between Cork and St. Petersburg. We, who have passed through the equally, or even more heavy emotional shock of the World Wars, have witnessed how they produced, in their aftermath,not only political and economic rebellions and revolutions, but rebellions affecting every aspect of moral, artistic and scientific life, from sex to semantics. If not outright rebellions or revolutionary, sharply critical attitudes could prevail even among persons of conventional life style. It thus becomes easier for us than for the Victorians and Wilhelminians to understand the state of mind that prevailed, especially among the younger generation, during and after the Napoleonic cataclysm. Thus the new mathematics of the period was only one aspect of that vigorous pioneering and rebellion that went on in almost all intellectual life in this period from 1789 to 1848, between the first and the third French Revolution. Think of the new ideas: in politics Republicanism and Carbonari, in economics the theories of Adam Smith and Ricardo and Fourier, in literature the modern novel with Dickens and Stendhal and in poetry the visions of Shelley, in theology the new critique of Strauss, the German philosophy from Kant to Hegel, the new linguistics of the Grimms, socialism and communism[3)] - and equally the spate of new ideas and theories in physics, chemistry, biology and geology. The gods themselves were challenged by invoking the spirits of Faust and Prometheus, and nationalism tended to replace the cosmopolitanism of the intellectual world in the previous century. If rebellions against the Holy Alliance and the Church were rampant, why not against the legend of creation, Newton's theory of light and even that pillar of sientific security, Euclid? Let us quote what Eric Hobsbawn, the English historian,

[3)] This trend in literature and arts is known as romanticism. It is preceded by a period often called classical. J.E. Hofmann calls the whole period from c. 1550-1700, baroque (in his "Geschichte der Mathematik"). There is no objection to call in mathematics the period from c. 1650-1700 classical (the "Hochbarock" and "Spätbarock" of Hofmann). But what is "baroque" in the mathematics of the scientific revolution except the periwigs of the mathematicians ? It is the same with the term romanticism. Cauchy's complex functions and Abel's elliptic functions are not more romantic than Lagrange's real functions and Legendre's elliptic integrals. What the mathematicians and the poets had in common was a critical attitude with respect to their predecessors as well as a growing national feeling and we therefore prefer to call the period from 1789-1840 the critical, or perhaps the national-critical period. But the mathematicians only looked forward, the romantics sometimes forward, sometimes backward. This relationship (or non-relationship) between romanticism and the new attitude in science deserves deeper study (e.g. the influence of Naturphilosophie).

has said about this exciting and creative intellectual life, this outburst of vitality:

"No one could fail to observe that the world was transformed more radically than ever before in this era. No thinking person could fail to be awed, shaken and mentally stimulated. It is hardly surprising that patterns of thought derived from the rapid social changes, the profound revolutions, the systematic replacements of customary or traditional institutions by radical rationalist innovations, should become acceptable... We know that the adaptation of revolutionary new lines of thought is normally prevented not by their intrinsic difficulty, but by their conflict with tacit assumptions about what is or is not "natural". It may take an age of profound transformation to nerve thinkers to make such decisions, and indeed imaginary or complex variables in mathematics, treated with puzzled caution in the eighteenth century, only came fully into their own after the Revolution."[4)]

It is easy to find other examples of mathematical fields, long neglected mainly because of the intensive cultivation of the calculus during the eighteenth century due to its interest in Newtonian mechanics. Projective geometry already goes back to Desargues and Maclaurin, the foundation of the calculus on the limit concept to D'Alembert. Even a non-euclidean geometry had appeared in Thomas Reid's theory of vision of 1764. Nobody in particular cared. It even needed the French Revolution to introduce a metric system of weights and measures, an idea already

[4)] E. Hobsbawn, The age of revolution 1789-1848, Mentor Book, New York, Toronto, 1962, p.345.

proposed by Stevin in 1585. Professor Wilder in such cases speaks of cultural stress and cultural lag.[5)]

Hobsbawn also points the many, now familiar, terms originating in this period 1789-1848, words we cannot do without, such as industry, factory, capitalism, railway (railroad in the USA), liberal, conservative (in the sense of party), nationality, engineer, liberalism. We can add mathematical terms as complex number and complex function, vector, determinant, potential, contour integral, projectivity, polarity, number congruence, lines of curvature, analytical geometry (in our present sense). Even the word scientist dates from this period, denoting a need felt at the time for a word expressing what had became a widely recognizable profession (Whewell, 1840). The term technology, introduced in 1769 by Beckmann, became only current in this period.

3 As we said, a good deal of the new mathematics originated at, or was directly influenced, by what was going on at the Polytechnique. Through the teachings and research of Monge, his colleagues and pupils Hachette, Biot, Malus, Dupin, Poncelet the whole aspect of geometry was changed; starting from what was Monge's new discipline, descriptive geometry. Here again we find one of these renovations of the type mentioned by Hobsbawn. Monge put orthogonal projection firmly on the mathematical map - but Dürer used it successfully as early as 1525, after which it was almost forgotten in favor of linear perspective. In Monge's school grew projective and analytic geometry, and by the application of calculus,

[5)] J.R.L. Wilder, Evolution of mathematical concepts. An elementary study, Wiley, New York, etc., 1968, passim.

differential geometry. This again fertilized the calculus itself, and with it mechanics. Fourier demonstrated the power of his trigonometric series in what may be called the opening up of mathematical physics; Cauchy developed the theory of complex functions and was one of the first to rescue the limit concept for foundation of the calculus. Lagrange and Laplace are also connected with the Polytechnique, and so is Poisson. A new type of textbooks appeared, result of classroom teaching, direct ancestor of our own college texts, such as Biot's and Hachette's books on conics and second degree surfaces, introducing the term *géométrie analytique*. But it did not all happen in Paris; Gergonne, in Montpellier, in 1810 began to publish the first periodical exclusively dedicated to the mathematical sciences, the "Annales de mathématiques pures et appliquées" (till 1832). And one of the sources from which sprang projective geometry with its concept of polar reciprocity. The many textbooks of Lacroix served as models for years to come.

Germany, in the time of Lessing and Goethe, had its own Enlightenment, but German middle classes being economically and politically weak, the stress here was on literature and philosophy, and not on the sciences. Germany had its Gauss, enthroned in Olympic isolation in his Göttingen observatory, like the equally Olympic Goethe in his Weimar *Kleinstaat*, both representing in a sense the transition from the old to the new on the highest level. It was the trauma of the Napoleonic invasion that woke Prussia up from its selfindulgence, and the university of Berlin, founded in 1810 under the influence of Wilhelm von Humboldt, became a model for the further development of the German university system, taking its place beside the older institutions like the ones at Göttingen and Königsberg. The earlier years at Berlin were strongly humanistic, reflecting the influence of Wilhelm von Humboldt (think of Ranke and

F.A. Wolff) and the still general weakness of economic development, but the mathematical sciences took a more leading position with the return of Wilhelm's brother Alexander from Paris, where he had mixed freely with the leading mathematicians. That happened in the 1820's. Then we find in Berlin Steiner and Dirichlet, representing novel approaches in geometry and analysis, while Crelle, the architect-engineer, begins his "Journal für die reine und angewandte Mathematik" in 1827, opening its pages for papers of the young and creative like Abel and Jacobi, incidentally the first Jew since the Middle Ages to occupy a leading position in mathematics, and in his case even a university position. The French Revolution also in this domain showed its emancipating influence. Germany begins its own path to mathematical glory, following in its own constructive way the lead of France and of Gauss.

4 In the new mathematics of the period of revolution the classical union of calculus and mechanics, typical of Lagrange and Laplace, is maintained, but supplemented by new and critical concepts. The way the eighteenth century worked with a calculus without satisfactory foundation, with infinite series without satisfactory study of convergence and with the "paradoxes of infinity" in general, was found highly unsatisfactory. With the new rigor came new criteria for the convergence of series and new understanding of such concepts as continuity and function. We think of Cauchy, Gauss, Bolzano, Abel, Fourier, Dirichlet.

We mentioned already the widening of the field of geometry. Started in France with Monge and his school, continued by Chasles, the Germans took over and developed affine, algebraic and more dimensional geometry, with Steiner, Grassmann, Plücker, von Staudt. Projective and more dimensional geometries can be seen as breaks with the Euclidean "paradigm".

Hardly any mathematical thinker in previous time had thought in terms of spaces of more than three dimensions, or, if he did, he certainly did not follow it up. This holds even more for non-euclidean geometry, embodying an idea so new that Gauss only discussed it in some letters to friends, but which, after an incubation period of two millennia, finally was boldly presented to a still sceptical public and by two men outside of the mathematical centers, one in Russia, the other in Hungary. Once its validity was recognized, it revealed itself as one of the most revolutionary, far reaching, discoveries in mathematics and far beyond it in science and philosophy.

Not only the mystery of the "metaphysics" of the calculus was dispelled, but also that of the imaginary and complex numbers. For centuries a mystique had surrounded $\sqrt{-1}$, and though Euler and others had shown by amazing results that operating with it "worked", only when Wessel, Argand, Gauss, Hamilton established their geometric-algebraic interpretations did the imaginary lose its position as a bastard in the respectable mathematical family. With this recognition came the theory of complex functions developed by Cauchy and anticipated by Gauss.

Once the "legitimacy" of the complex number was recognized, the door was opened for its generalization. With the quaternions and other hypercomplex numbers came the concept of vector and of higher direct quantities. We think of Hamilton, Grassmann and others. Here again we discover how concepts, lying dormant for a long time, came to new life in this revolutionary period. Operations with geometrical quantities instead of numbers or letters was, as we know, already suggested by Leibniz.

Just as Leibniz' ideas about a calculus of direct quantities, so Lagrange's ideas on the solvability of algebraic equations were now seriously taken up and further developed. This concerned the ancient problem of why fifth and higher degree equations could not be solved in radicals the way equations of lower degree could be solved. Through the work of Ruffini and Abel this led to one of the most fertile of the new mathematical ideas, Galois' theory of groups.

Elliptic integrals were long known and Legendre had written an extensive classification of these quantities and their transformations in 1811, continued in 1826. But the startling discovery of elliptic functions as doubly periodic inversions of these integrals was made by younger men, by Abel and Jacobi. This widened the bounds of function theory enormously, not only by the free use made of the complex domain, but also by the introduction of theta functions and Abelian integrals.

Jacobi also introduced determinants, as Cauchy introduced matrices. With vectors, quaternions and groups a new algebra was born, a radical departure from the age old identification of algebra with the theory of equations of different degrees. But here new ideas also came out of England in the work of DeMorgan, Peacock and Boole, steps leading to the axiomatization of algebra and even of logic.

5 This brings us to Great Britain, the European country that dedicated itself to fight the French Revolution and Bonaparte, and thus was most inclined to withstand the intellectual and political influence of this French Revolution and this Bonaparte. The opposition goes back to the whole of the eighteenth century and its repeated British-French wars. A bad period for the adaption, or even the serious study, of continental ideas, the more so since the native sons with scientific and engineering interests were absorbed, directly or indirectly, in research and construction related to the industrial revolution. Continental mathematics was not very welcome.Not even the decimal division of weights and measures was

allowed to enter Great Britain. Yet, "subversive" penetration of the ideas of the French Revolution did exist - we may think of Goodwin and Mary Woolstonecraft. In mathematics it came, in and around 1816, in the form of the Cambridge Analytical Society and its young members Babbage, Herschel and Peacock. They began to propagate the continental calculus, the "d-notation" instead of Newton's fluxions and translated a book of Lacroix. Babbage would soon start on his mechanical computer.

Another influence in the same direction was the creative study of Laplace's "Mécanique céleste" by Hamilton and by Green, leading Green to the mathematical theory of electricity and Hamilton (in Ireland) to his remarkable work on optics and mechanics, which in its turn led to the Hamilton-Jacobi theory and its far going consequences lasting till the present day.

We have already mentioned the algebraic work done in England without much influence from abroad - this therefore a native contribution, as was Boole's (also in Ireland) creation of mathematical logic.

We see that mathematical research, as well as teaching, was spreading outside of the West European heartlands. In the USA we find Bowditch translating the "Mécanique céleste", and Farrar at Harvard introducing continental calculus, like the young men at Cambridge, French texts in translation. From Skandinavia came Abel, from Russia Ostrogradsky and Lobatchevsky, from Hungary Bolyai. Attempts in Mexico to introduce continental calculus at the newly founded mining institute (the Minería, 1792) were eventually frustrated by the War for Independence.

New perspectives were opened in education, in the first place for the middle classes. The national states needed professionals for their growing bureaucracy, schools and industry. The industrial revolution spread over the continent and needed engineers. University systems were modernized and technical institutions founded, professionalism and specialization encouraged. Monge and Steiner were geometers, Laplace and Dirichlet analysts, Peacock an algebrist, Boole a mathematical logician. The road was open, in several of the leading countries, for talent, especially young talent. In 1814, when Comte entered the Polytechnique, Cauchy, among the instructors, was 26 years of age, Arago 28, Poisson 33, Poinsot 37, Ampere 39, several had seen many years of service. Hachette, at 45, already had pupils as colleagues. "L'Empire n'est pas seulement le temps des jeunes généraux, c'est aussi celui des jeunes professeurs", writes Comte's biographer at the Polytechnique. Only Monge, Lagrange (already dead) and Laplace belonged to the older generation.

Since specialization had set in, the gap between pure and applied mathematics was widening, but not all links disappeared. In France the connection remained strong, as in the case of Cauchy and Poisson. Mathematicians like Dupin and Poncelet showed a deep interest in politics and in industry. Rodriguez was a follower of Saint-Simon. Gauss, in Germany, was of course master in both fields. We think of the wellknown interchange of opinions between Jacobi and Poisson: is the ultimate goal of mathematics utility or the honor of the human mind? Yet it was Jacobi who was interested in establishing a Polytechnique in Berlin.

This was also the age of periodicals purely devoted to mathematics. We have mentioned Gergonne's Annales and Crelle's Journal. After Gergonne's Annales had expired Liouville started, in 1836, his "Journal de Mathématiques pures et appliquées". Pure and applied indeed, but Crelle's Journal soon was nicknamed "Journal für die reine unangewandte Mathematik". In 1839 the Cambridge Mathematical Journal was founded.

A word about the historiography of mathematics in this period. The

French Enlightenment had brought the first readable history of mathematics, a splendid narrative quite different from the stale catalogs of names and titles that had appeared before. It was Montucla's "Histoire des mathématiques" (2 vols, 1756, 4 vols 1799-1802, the last volumes completed by Lalande). Some historics of minor importance appeared around the turn of the century, those by Bossut (1802), well written, and Kästner (1797-1800), more a descriptive catalog. The new age went in for specialization. Already in 1797-99 Cossali had published his 2 volumes on the origins of algebra in Italy, with a light patriotic touch. Far more patriotic was Libri's history of mathematics in Italy (1830-41). Chasles' "Aperçu historique" of 1837 dealt with the progress of geometry through the centuries, integrating ancient and modern results into a living pattern, the first book on the history of an important field of mathematics written by a creative mathematician. Nesselman's "Algebra der Griechen" was the solid work of a pupil and later colleague of Jacobi.

This was also the period in which the horizon was widened far beyond the limits of the old classical-European world. The search for information in connection with markets and imperial expansion brought scholars to explore the East. This brought Rosen, Woepcke and the Sédillots to the study of Arabic mathematics, Colebrooke and Strachey to the mathematics of the Hindus. With Biot and Wylie begins the modern study of Chinese mathematics. But despite the beginning flowering of Egyptology and Assyriology the discovery of the mathematical treasures hidden in hieroglyphics and cuneiform was still in the future. The total harvest in our period remained small, and only during the second part of the century the new and fertile period in the historiography of mathematics is opened.[6)]

6) See my article in NTM (cf. Select Bibliography).

ORIGINS OF THE PROGRAM OF "ARITHMETIZATION OF MATHEMATICS"

Hans Niels Jahnke and Michael Otte

1. Introduction

Curiously enough, mathematics and its historiography are rather acutely conscious of the fact that the turn from the 18th to the 19th century marks a decisive turning point full of consequences in the development of science. Contemporaries in the 18th century believed that mathematics had come to an end of its growth. "A great upheaval in the sciences is imminent. In view of the present aspiration of the great minds, I should almost like to claim that there will not be three great mathematicians in Europe within a century. This science will suddenly remain fixed to the spot where the Bernoullis, Euler, Maupertius, Clairaut, Fontaine, d'Alembert, and Lagrange have left it." (Diderot 1754, p. 31). Similar statements have come down to us from Lagrange.

It is a well-known fact that Diderot's and Lagrange's fears have not come true; rather, the "great upheaval in the sciences" predicted by Diderot seized mathematics as well, and led to new developments of method and object not anticipated. The new style of mathematics, which began to emerge at the turn to the 19th century, is seen, as most historians of mathematics agree, first of all in the tendency towards rigorous proof, and in a more careful elaboration of the foundations and definitions of mathematics. Analysis sees a foundation of its methods, the nucleus of which is described as arithmetization. In retrospect, Felix Klein wrote in 1895: "The spirit in which modern mathematics was born, however, is quite another one. Starting from the observation of nature, and aimed at explaining nature, it has placed topmost a philosophical principle, that of continuity. This applies to the great pioneers, for Newton and Leibniz, it applies to the whole of the 18th century, which, for the development of mathematics, has really been a century of discoveries. It is only gradually that rigorous criticism emerges, which enquires after the consistency of these bold developments - something like a re-establishment of ordered administration after a long campaign of conquest. This is the age of Gauss and Abel, of Cauchy and Dirichlet ... hence the demand for exclusively arithmetical proof." (Klein 1895, p. 143/144)

This summary by Klein represents a view of the development of mathematics in the 19th century current far and wide to this day. It entails, however, some difficulties and problems.

The first problem is of immanent order and concerns the opposition established by Klein between mathematical discovery in the 18th century, and the foundation resp. codification of mathematics in the 19th century. Is it true that this codification has nothing to do with the development of new knowledge, should the new foundation of mathematics have no productive function at all? Does this separation between development and foundation really apply to the mathematics of the 19th century, which did show the marks of a historically unprecedented productivity? The second problem lies in the question whether, and how, this recourse of mathematics to its own foundations is connected with the fact that the sciences, in the 19th century, are, for the first time, directed towards application on a broader and socially pertinent scope, and that its entire social and institutional basis is subject to a fundamental change of structure. In our opinion, this question should be made the starting point for any analysis aiming to study the historically unique character of mathematics' development during the first half of the 19th century.

These questions sketch a program the implementation of which will require studies of different types, and which cannot be realized, either, in the domain of mathematics history alone. The connection between the sciences and their applications on one hand, and the social environment on the other, cannot be studied in an isolated fashion according to individual disciplines. That which has been described by others as "Finalisierung" (see Böhme et al. 1973) cannot be a statement pertaining to an individual science, but rather represents a (historical) characterization of the totality of the science system. In the period under consideration, i.e. the turn to the 19th century, this idea, i.e. that the relationship between theory and practice can only be discussed referring to the totality of the sciences, was an essential point of debate. The attacks, in particular those of German Neo-Humanism, against the "utilitarian thinking of the Enlightenment", were not directed against the orientation of the sciences toward application, but rather against a too narrow understanding of that which is to be understood by application of science. Neo-Humanism is concerned with developing conceptions for the development and the application of science, which are more appropriate to the social character of science, resp. which foster this process of socialization. The same intention seems to underlie Saint-Simons famous remark which has been taken from quite another context of discussion: "The philosophy of the 18th century was revolutionary; that of the 19th century is called upon to organize."

To explain autonomy *and* dependence of the sciences simultaneously, in our opinion, thus seems to be possible only if the totality of the sciences is considered an essential element of the context of the explanation. This level of "totality" is not identical to the direct relationships between the various disciplines, but represents, in itself, a new level. With regard to the institutional and material basis of the sciences, this is expressed in the requirement to view the sciences as components of a total system, which might tentatively be designed by the term of "superstructure". This "superstructure" has its specific organizational and material foundations and is determined by the latter. An important point of investigation, for instance, would be to study the superstructure's technologies, such as printing technologies, experimental techniques, present-day media, etc.

On the other hand, the element of the totality of the sciences is present on the level of knowledge as well in the shape of philosophy, or "meta-knowledge". It is only in this medium that "boundary concepts" are generated, "which do not only orientate scientific research in a fundamental way, because they 1) are immediately accessible to content-related intuitions, 2) possess a methodological constructivity, and 3) are of theoretical fundamentality, but are fundamental concepts as well, i.e. fundamental concepts in the sense of linking scientific research to the other dimensions of human orientation." (AG Mathematiklehrerbildung 1981, p. 158)

This background makes plausible that the connection between science and education is of such extraordinary importance for investigating the development of science at the outset of the 19th century. To apply science is not merely to provide concrete knowledge for the solution of concrete problems. Rather, the function of science to provide general orientations is essential for developing an active-practical behaviour toward real life. "If one tries for a better understanding of the connection between education and science, it must be realized that education establishes above all the individual's conscious and also unconscious relationships to knowledge, produces ideas of the unity and coherence of knowledge, and furthers the methodolisation of knowledge. Education is connected in an essential sense with the establishment of scientifically general, supradisciplinary concepts. ... Education is thus revealed as an important field in which the very connection between epistemology and social theory becomes operative, which we supposed above to be characteristic of the newly emerging rationality type." (Jahnke/Otte/Schminnes 1981, p. xx-xxi)

Indeed, social theory, personality theory, and theory of science have entered into close combination in the early 19th century pedagogical literature of Germany in particular. A further characteristic is that the pedagogical literature cannot be clearly separated from the philosophical. Methodological and social aspects of the development of science are considered closely connected. Contemporary French positivism, too, is strongly rooted in pedagogical thought (see Cassirer 1957, p. 17ff).

In studies pertaining to the methodological change at the turn to the 19th century, history of science thus must incorporate that type of literature in which this way of thinking emerged. Hence, in the German case, for instance, the essential comprehensive mathematical teaching and textbook literature as well as, say, the "Programmschriften" of the

Gymnasia must be taken into consideration. To show the connection between the thought reflected in this type of literature, on the one hand, and the actual academic research on the other, is an important requirement asked of history of science. A first essential step could focus on such authors who, in their time, consciously assimilated the literature mentioned above, and who, in addition to that, had an impact on the most advanced level of scientific discussion. With regard to mathematics, Bernard Bolzano is undoubtedly such a person; others, however, might also be listed.

To be concerned with social history of science cannot mean simply to reveal "social influences" exerted on the sciences. Scientific knowledge is not socially constituted for the mere reason that the sciences represent an activity of human beings who act and communicate. Such a vague use of the term "social" would disregard the specific nature of the scientific field, "which is the specifity of the politics of truth in our society". (Foucault 1977)

Hence, social history of science also implies close study of the subject matter and applications of scientific theories. In this respect, the social history of science is no opposite to the history of ideas. The philosophical understanding of the object field of scientific theories, conceptions of what is usually called the subject matter of a scientific theory, provides an essential yardstick for the progress of the sciences itself, and this not only with regard to the contents of knowledge, but also with regard to the very social and institutional foundations of the sciences. Any increase in differentiation of the object field will be accompanied by an increase in differentiation and explanation of the scientific system in its social, institutional, methodological, and literary components. In this sense, the following will attempt to describe some aspects of the change undergone by the methodological self-understanding of mathematics in the early 19th century.

2. Arithmetization of Mathematics

An attempt for better comprehending the change in mathematics' understanding of its object field, which began to take place towards the turn to the 19th century, first of all requires tackling the problem and the context of the "arithmetization of mathematics". It must be said at once that this arithmetization was not only, as is often supposed, a matter of founding infinitesimal calculus anew, but rather of reshaping and reformulating mathematics as a whole. The nucleus of that which will be termed arithmetization here, might be approximately described as follows: During the 18th century, numbers, in their inseparable linkage to the quantity concept, represented the actual object field of mathematics, and algebra, and the symbolic calculi of mathematics were regarded merely as a language permitting an easy and suggestive manner of representing relationships between numbers or quantities. This status became precisely the reverse in the 19th century. Algebra was now to directly include the actual mathematical relationships, which constitute the subject matter under study, while arithmetics, for its part, became the language of algebra resp. of the entire mathematics, by means of which, and in which, all mathematical facts must ultimately be expressible. This process of arithmetization finally culminated, towards the close of the century, in the fact that the consistency of mathematics was reduced to the consistency of arithmetics, raising arithmetics to the position of foundational science proper of mathematics. Hilbert's program, which attempted to reduce the entire mathematics to finitist combinations of signs resp. numbers, is but a pointedly formulated version of these efforts, which, eventually, led to Goedel arithmetizing the logical system of the "principia mathematica".

Arithmetics as a foundational science of mathematics does not mean that arithmetics constitutes the actual subject matter of mathematics. Rather, numbers are no longer interpreted as objects, but as pure symbols, as "marks", as a means of objectifying mathematical thought - i.e. as a language. This will be exemplified by quoting one of Helmholtz' remarks, who says in his fundamental essay "Zählen und Messen, erkenntnistheoretisch betrachtet" (Counting and Measuring from an Epistemological Point of View, 1887): "I consider arithmetics, or the theory of pure numbers, as a method based on purely psychological facts, which serves to teach the consistent application of a system of signs (i.e. numbers) of unlimited extent and unlimited opportunities of sophistication. In particular, arithmetics explores the question which different ways of combining these signs (calculating operations) will lead to the same final result. ... Besides the proof thus furnished of the inner consistency of our thoughts, such a method would of course at first be a mere play of our acumen with imagined objects, ... if it did not permit such extremely useful applications." (Helmholtz 1887, p. 303/304). In connection with non-Euclidean geometry, Gauss wrote to Bessel on April 9th, 1830: "According to my innermost belief, the status of space theory with regard to our a priori knowledge is quite different from that of pure quantity theory; our knowledge of the former must do quite without that utter conviction of its necessity (that is, of its absolute truth as well) which is proper to the latter; we must modestly admit that, if the number is merely the product of our mind, space has a reality outside our mind as well, to which we cannot a priori prescribe its laws." (quoted after Becker 1975, p. 179)

There was the widespread idea that number theory was the purest expression of the laws ruling our thoughts. Hence it seems to be no accident, even if the causes, as a whole, may have been much more complex, that number theory saw such a flourishing growth towards the beginning of the 19th century. Crelle's following statement is typical: "This theory of numbers now is, at least in its further extension, a n e w e r branch of mathematics, comparable to differential and integral calculus. Only the first traces of it are to be found with the ancients. Its development did not begin until Fermat's time, and it has reached its present scope but recently; particularly due to the efforts of Euler, Lagrange, Gauss, Legendre; later to those of Jacobi, Dirichlet, etc. No matter how novel it is, however, it has made an unprecedented advance. It has grown to a large amount of theorems, and is extending daily." (Crelle 1845, p. V)

3. On Hamilton's Number Concept

The conception saying that numbers are signs rather than objects by no means entailed that mathematics became a science removed from real life. On the contrary, the relative separation of the sign level from applications expressed by this conception fostered the relation of mathematics to reality. This can be shown in an exemplary fashion in Hamilton's efforts to establish algebra as a "science of pure time". Hamilton frequently is considered an adherent of Kant. As far as this may be correct, it does not refer so much to Kantian epistemology, however, but rather to the dynamistic (anti-atomistic) conceptions developed by Kant in one of his early stages. This makes Hamilton just one representative of a host of British scientists, who were influenced by the German Philosophy of Nature, by Kant and Schelling, in their efforts to come to an appropriate understanding of physical and chemical processes. Coleridge and

Humphrey Davy are the most influential proponents of dynamical philosophy in England. (see Williams 1965, p. 63ff.)

This connection with the German philosophy of nature already shows that Hamiltons thoughts were strongly guided by physical-dynamical intuitions, from which he developed his conception of mathematics. Hence, algebra is founded by Hamilton under the perspective of a general hypothesis about the world's material structure. By his dynamical ideas, Hamilton is led to the belief that space and time are not independent of each other, but form a unity. Hamilton searches for a new relationship between geometry and algebra (see Hankins 1977, p. 178). These conceptions guided him for years in his search for the quaternion-calculus, which was to express this unity of space and time. Hamilton is anti-atomist, he aspires at supplanting the theory of atoms by a theory based on the "forces" and "energies" acting in space and time. In June 1834, he had a decisive meeting with Faraday in Dublin, which encouraged him to pursue his ideas on this matter. In the last instance, highly general physical ideas in the sense of an intended reference to applications entered his considerations on the foundation of mathematics, and of algebra in particular.

In his fundamental work "Theory of Conjugate Functions, or Algebraic Couples; with a Preliminary and Elementary Essay on Algebra as the Science of Pure Time" (1837), in which Hamilton develops his conception of algebra as the "science of pure time", he starts from the assumption that there are three possibilities of conceiving algebra: a "practical" one, which considers algebra as being purely instrumental, a "philological" one, which sees in algebra a pure calculus, oriented toward a symmetry of expressions, and a "theoretical" one. It is the theoretical conception he seeks to develop, and it consists precisely of the attempt to identify an object of algebra. After having shown that neither the appearance of negative nor that of imaginary numbers can be founded on the quantity concept, he develops a conception of his own which is based on the concept of "progression" or of "order in time". Hamilton thus supposes that the time axis is given, together with a relation of order between the point situated on this axis. Then he proceeds to consider pairs of moments (points in time), which can be compared with regard to the relation of order, i.e. with regard to "before" and "afterward", and finally assigns, to each pair, a "step", i.e. a translation which translates the earlier point in time into the later one. Subsequently, he develops the laws of calculating by means of such steps, and finally he is able to quite generally introduce (real) numbers as "quotients" of two of these translations, respectively. It is well known that Hamilton, in his presentation, anticipated much of the foundation of the number concept developed later; he also gets very close to Dedekind's intersection axiom in showing that the "quotient" of two steps will not always yield a rational number, using a construction analogous to Dedekind's intersection axiom in doing so. The core of Hamilton's approach is that numbers are being introduced, by means of these definitions, as pure numbers resp. marks for the relations between steps, thus converting algebra to an aprioristic, completely certain science. He expresses this quite unequivocally in an unpublished manuscript quoted by Hankins (1976) and Mathews (1978): "In all Mathematical Science we consider and compare relations. In algebra, the relations which we first consider and compare, are relations between successive states of some changing thing or thought. And numbers are the names or nouns of algebra: marks or signs, by which one of these successive states may be remembered and distinguished from another ... . Relations between successive thoughts thus viewed as successive states of one more general and changing thought are the primary relations of algebra... . For with Time and Space we connect all continuous change, and by symbols of Time and Space we reason on and realise progression. Our marks of temporal and local site, our then and there are at once signs and instruments of that transformation by which thoughts become things, and spirit puts on body, and the act and passion of mind seem clothed with an outward existence, and we behold ourselves from afar. And such a transformation there is when in Algebra we contemplate the change of our own thoughts as if it were the progression of some foreign thing and introduce numbers as the marks or signs to denote place in that progression." (Quoted after Mathews 1978, p. 188)

The meaning of Hamilton's foundation of mathematics will only be understood if there is a careful distinction between a concept's simulative function, i.e. the production and variation of sign models, and its explorative function, i.e. the intended reference to the object, which continuously reflects congruence and difference between the concept and the object. Numbers, as objectified helps for acts of thought, serve to simulate. Numbers are the matter used to produce symbolic models. The explorative content of theory, however, lies in its intended reference to physical applications, which is conveyed by the concept of "progression" or "order in time". Subjectivation of the sign level, conceiving of numbers as of pure symbols of thinking acts will not retain a rational meaning unless it is regulated by conceptual generalizations, which refer to the objectified field of application to a far more comprehensive and extensive degree than was hitherto the case. The simulative and the explorative function of the concept are only conveyed within, and by means of, the activity of cognition, and communication. The fact that activity becomes the point of reference for the understanding of scientific generalization is the decisive achievement of 19th century epistemology.

4. Gauss and the "Metaphysics of Mathematics"

Gauss' dislike for most philosophical schools has led many authors to believe that he considered any treatment of questions of philosophy of science irrelevant. Documentary evidence, however, shows that the opposite is true, and that Gauss thought about philosophy of science and mathematics in a very profound and intense way. Those philosophical manuscripts of his which have been handed down to us contain, despite their cursory character, the nucleus of a very far-reaching meta-mathematical conception which helps us to decipher important aspects of the self-understanding of mathematics during the early 19th century.

Among these philosophical manuscripts, we also count the well-known "2. Ankündigung der Theorie der biquadratischen Reste" (Second Announcement of the Theory of Bi-Quadratic Residues) published in 1831, in which Gauss founded the introduction and admissibility of complex numbers. A closer look at this text will be rewarding, as it can be shown that Gauss did not consider the "illustration" of complex numbers in the plane the essential feature. Gauss writes: "Positive and negative numbers will only lend themselves to application, when that which has been counted has an opposite, which, if thought combined with the former, amounts to annihilation. Upon closer look, this prerequisite will be given only in cases where, rather than substances (objects which can be thought apart), relations between two objects, respectively, are that which has been counted. The postulate is that these objects are ordered in a series according to a certain manner, e.g. A, B, C, D..., and that the relation of A to B can be considered equal to that of B to C. The concept of opposition, here, requires nothing more than an exchange of the relation's members, so that, if the relation (or the transition) from A to B is considered to be +1, the relation from B to A must be represented by -1. Insofar as such a series is unlimited to both sides, each real whole number thus represents the relation of one member arbitrarily selected at the beginning to one very distinct member of the series." (Gauss 1831, p. 175/176) This, in all its briefness, is a "definition" of whole numbers which corresponds to that which Hamilton comprehensively developed in his algebra as a science of pure time. Also remarkable is the insight that it is already the negative numbers which require that the concept be conceived of as the designation of a relation rather than as a name of a substance.

For the complex number, Gauss continues as follows: "If, however, the objects are of such kind that they cannot be ordered into a single, if unlimited series, but only into series of series, or, what amounts to the same, if they form a variety of two dimensions and if a relationship holds between the relations of one series to another, or between the tranistions from one to the other, which is similar to the transitions already mentioned from one member of a series to another belonging to the same series, measuring the transition from one member of the system to another will require, in addition to the units +1 and -1 already noted, two others, which are also opposed to each other, +i and -i. Evidently, this requires the additional postulate that the unit i always denotes the transition from a given member of a series to a determinate member of the immediately adjoining series. In this manner, it will be possible to order the system in a double way into series of series. The mathematician abstracts entirely from the quality of the objects and from the content of their relations; he is only concerned with counting and comparing their relations among themselves: Insofar he is entitled, just as he assigns similarity to the relations designed by +1 and -1, seen as such, to extend this similarity to all four elements +1, -1, +i, and -i." (Gauss 1831, p. 176)

Gauss begins by developing a general concept circumscribing an intended field of application characterized by a high degree of generality, and by an extensive scope. Of course, such a description makes sense only if there is an activity which refers to this field of application, and is precisely formulated with regard to its content, and if he sees how this concept refers to the theory of the functions of two variables, resp. to number theory in this context. Only then Gauss continues and provides the well-known illustration of this notion on the Euclidean plane, emphasizing the symbolic, simulative character of this illustration by his choice of terminology.

In his letters, Gauss repeatedly stated that the illustration in the Euclidean plane does not represent the essence of his foundation of the concept of the complex number. To Drobisch, he wrote: "The representation of the imaginary quantities, however, by means of the points in the plane, is not their very essence, which must be conceived of as being higher and more general, but rather the example of their application purest to man, or perhaps even the only really pure example of their application." To Hansen, he wrote: "The true meaning of $\sqrt{-1}$ is very vivid in my mind, but it will be very difficult to grasp with words, which will always give but a vague image floating in the air... ." (Both quotes after Schlesinger 1912, p. 56)

It can be shown that Gauss is concerned with thoughts on the "ontology" of mathematics here, which go far beyond the occasion of introducing complex numbers. Indeed, there is a short manuscript of one and a half pages among his work, which was given the title "Zur Metaphysik der Mathematik" (On the Metaphysics of Mathematics) by the editors, and which was probably written in 1825 or 1826. In this manuscript, Gauss begins with the general question: Which is the essential prerequisite permitting that a linkage of concepts be thought as referring to a quantity? This is a question enquiring after the essential prerequisite for the fact that a theory pertaining to a subject matter field can be mathematized.

Gauss gives a quite universal answer to this question, saying that mathematics is in the most general sense the science of relations, abstracting from all content of these relations. He explains this general statement by placing a drawing of the whole number plane on the margin, saying that points should be conceived of as objects, and transitions as relations, just as in this drawing. The general notion of things, in which each has a relation of inequality only with regards to two other things, is that of points on a line. If one of these points can have a relation to more than two others, we must picture these as points in a plane, which are connected by lines. If, however, study shall be possible here, it can only concern those points entertaining a mutual relationship within three others, a relation existing between the relations (Gauss' Werke X/1, p. 396/397)

Hence, mathematical concepts do not represent things, but relations between things. Cassirer has provided a detailed analysis of this transition from thinking in objects to relational thinking (Cassirer 1910, 1976). Beyond that, however, two other elements of the text quoted are important. First, this text indicates the prominent role of the discrete (of whole numbers) for understanding relational thinking, and second, it emphasizes that the quantity concept was no longer sufficient to characterize the object field of mathematics.

With regard to the historical contexts of this note by Gauss, it will first of all have to be realized that there were decisive efforts undertaken at the turn to the 19th century to apply mathematics in several empirical object fields outside of mechanics, geometry, or astronomy. Most prominent among these are the theory of heat (Fourier), the theory of electricity, and the theory of magnetism.

Besides these efforts at mathematization, which were ultimately successful, there were efforts as well to apply mathematics to the social sciences, resp. to psychology (e.g. Herbart). The attempts to use mathematical methods in chemistry, too, were fundamental and more far-reaching in their intentions than was proved possible in the end. Gauss himself was deeply involved in these attempts; the measuring problem linked to the application of mathematics is an essential component of his scientific biography. This refers not only to his substantial geodetic surveying, but also to his successful efforts of many years at surveying earth's magnetic field. Both series of surveys are not only scientific masterpieces in a cognitive sense, but also represent an important organizational achievement. In retrospect, it seems obvious that Gauss was very interested in theoretical concepts and ideas related to the problem of measurement.

On the background of the problems raised by the task of mathematizing and quantifying fields of experience hitherto not mathematized, Gauss' above position, which was not only his own, but widespread in his time, can be interpreted convincingly. In epistemological terms, quantifying an object field (the transition from the qualitative to the quantitative) is not to be imagined as consisting of a) previous identification of the decisive quantities, b) development of methods to measure them, and c) final empirical discovery of the important natural laws. Rather, the determination and definition of the relevant quantities is itself dependent on a previous knowledge of the relationship, for which these quantities are sufficient. To quantify a field of experience thus is to intervene into this field of experience, and to change it. In the philosophy of science, these problems are well known under the heading of "theory-loadedness of empirical terms".

In order to show what is the point without getting lost in technicalities we shall quote a statement of G. Böhme on the connection between scientific experience and everyday experience which expresses the same problem. "A fifth characteristic of everday life experience is that all qualities are polarized and often show an inner structure of harmony resp. disharmony. There are heavy and light things, there is heat and cold, there are high and low tones. The actual phenomena will always be determined by this polarity's span. Science, however, will tolerate polarities almost nowhere. Its objective is to achieve general comparability of phenomena in one field. There are no more heavy and light things, but merely more or less heavy ones. This "linearization" is the first step to quantification." (Böhme 1979, 126)

Compare this description of how everyday life qualities are linearized in relations (and thus mathematized) to A.N. Whiteheads following argument: "The whole difference between the older and the newer mathematics lies in the fact that vague half-metaphorical terms like "gradually" are no longer tolerated in its exact statements... . Of two numbers one can be greater or less than the other; and one can be such and such a multiple of the other; but there is no relation of 'graduality' between two numbers, and hence the term is inadmissible... . In working our way towards the precise definition of continuity (as applied to functions) let us consider more closely the statement that there is no relation of 'graduality' between numbers. It may be asked, cannot one number be only slightly greater than another number, or in other words, cannot the difference between two numbers be small? The whole point is that in the abstract, apart from some arbitrarily assumed application, there is no such thing as a great or a small number. A million miles is a small number of miles for an astronomer investigating the fixed stars, but a million pounds is a large yearly income... . Our task therefore is to define continuity without any mention of a 'small' or 'gradual' change in value of the function. (Whitehead 1961, p. 115-117)

The elimination of everyday life elements of description from scientific representation leads to the conception saying that scientific, mathematical concepts do no longer reflect things, but rather relations between things. The prominent role of arithmetics, of the discrete, is based upon this transition to relational thinking. We believe that the distinction between numbers and quantities is not trivial in this sense. As Bateson remarks: "It is impossible, in principal, to explain any pattern by invoking a single quantity. But note that a ratio between two quantities is already the beginning of pattern. In other words, quantity and pattern ... do not readily fit together in the same thinking." (Bateson 1979, p. 53). At the same time, and for the same reason, the quantity concept becomes obsolete, as quantities can no longer be empirically demonstrated, but rather as their definition results from the relational pattern, which they belong to. In addition to this, numbers are more suited to bring forward the procedural aspects of the generation of knowledge. The program of "Arithmetization of Mathematics" marks therefore not only the transition to relational thinking but at the same time gives the procedural aspect of knowledge a more prominent role. A theoretical concept not only embodies relations but simultaneously becomes a scheme of action in a new way.

The vicious circle of the relations determining the quantities, and the quantities determining the relations, can only be dissolved into a process in time.

Within this framework, the function concept, of course, is entitled to a prominent position. It becomes important as empirical research is now examining complex natural phenomena which can no longer be modelled such as to achive correspondence between a definitely given parameter, and a definitely given result. This conception which may be characterized as "theory of the one-factor-experiment" is no longer adequate to the situation. Rather, it is a question of various parameters varying against each other. Of course, a changed attitude towards the experiment results from the more complex understanding of the processes. It is no longer the single experiment which will serve to clarify a natural phenomenon. Rather, the transition to sequences and series of experiments, that is, to the establishment of an experimental practice, is the essential methodological characteristic of this process. In our opinion, the emergence of the general function concept at the turn to the 19th century does not just represent the development of a new concept. Rather, the function concept represents a fundamentally new model of scientific generalization.

5. Resume

As measured against the program sketched in the introduction, the above considerations have only yielded a contribution towards clarifying a detail. It has been attempted to show that the program of arithmetizing mathematics can be understood as a response to the changed relationship between mathematics and the empirical sciences. In that, we will have to stress that it is not so much a matter of the relationship between mathematics and any one specific discipline, but rather that the empirical sciences in their totality have created a new situation for mathematics. The new conception of the relationships between science and experience, science and society, science and progress, according to which it is considered useful to mathematize, say, psychology, has an impact on mathematics, and entails far-reaching consequences for the latter. This results in another level coming into play, which has played no explicit part in the above lines, but ought to be represented in the same manner, in case we intend to clarify the phenomenon of mathematics' arithmetization: the level of philosophy. We shall confine ourselves to the proposition that the early history of the program to arithmetize mathematics could also be written as a history of Kant's philosophy of mathematics, and the criticism advanced against it by the mathematics of the early 19th century.

According to our introductory considerations, philosophy, in the early 19th century, is deeply linked to the education problem. It is very revealing that there is, just at this time, an author in Germany, Martin Ohm, the brother of the physicist Georg Simon Ohm, who presented a draft of "a perfectly consistent system of mathematics" (Ohm 1822) explicitly based on pedagogical ideas. Judging by its inherent logic, this work can probably be said to be the most developed attempt to explain the program of arithmetizing mathematics (see Bekemeier 1980). In this respect, Ohm was recognized by Hamilton and Bolzano. This, however, creates a new outlook. The most current view of a connection between science and education, which simply regards education as a reproduction of the subject matter, will presumably not hold water. Rather, education is shown to be a field and an activity having a most essential part in founding science.

References:

AG Mathematiklehrerbildung: Perspektiven für die Ausbildung der Mathematiklehrer. Untersuchungen zum Mathematikunterricht 2, Köln 1981

Bateson,G.: Mind and Nature, New York 1979

Becker,O.: Grundlagen der Mathematik in geschichtlicher Entwicklung, Frankfurt/M. 1975

Bekemeier,B.: Zum Zusammenhang von Wissenschaft und Bildung am Beispiel des Mathematikers und Lehrbuchautors Martin Ohm. Materialien und Studien des IDM, Bd. 20, Bielefeld 1980

Böhme,G., van den Daele, W., Krohn, W.: Die Finalisierung der Wissenschaft, 1973, hier nach: W. Diederich (Hrsg.), Theorien der Wissenschaftsgeschichte, Frankfurt/M. 1974

Böhme,G.: Verwissenschaftlichung der Erfahrung. Wissenschaftsdidaktische Konsequenzen, in: Böhme/Engelhardt: Entfremdete Wissenschaft, Frankfurt 1979, p. 114-136

Cassirer,E.: Das Erkenntnisproblem in der Philosophie und Wissenschaft der neueren Zeit, 2. Band, Darmstadt 1974

Cassirer,E.: Substanzbegriff und Funktionsbegriff, 1910,

hier: Darmstadt 1976

Crelle,A.L.: Encyklopädische Darstellung der Theorie der Zahlen, Bd. 1, Berlin 1845

Diderot,D.: Zur Interpretation der Natur, 1754, hier nach der Ausgabe Reclam Leipzig 1976

Foucault,M.: The Political Function of the Intellectual, in: Radical Philosophy 17 (1977), p. 12-14

Gauss,C.F.: Fragen zur Metaphysik der Mathematik. In: Werke X/1, p. 396/397

Gauss,C.F.: Theoria Residuorum Biquadraticorum. Commentatio secunda; Göttingische gelehrte Anzeigen 1831, April 23, Werke II, p. 169-178

Hamilton,W.R.: Theory of conjugate functions, or algebraic couples; with a preliminary and elementary essay on algebra as the science of pure time, 1837, in: W.R. Hamilton: The Mathematical Papers, vol. III, p. 3-96

Hankins,Th.L.: Algebra as pure time: William Rowan Hamilton and the foundation of algebra, in: Macam/Turnbull (ed.): Motion and Time, Space and Matter, Chapter 12

Hankins,Th.L.: Triplets and Triads: Sir William Rowan Hamilton on the Metaphysics of Mathematics, ISIS 68(1977), p. 175-193

Helmholtz,H.v.: Zählen und Messen, erkenntnistheoretisch betrachtet, 1887, in: Hörz/Wollgast (Hrsg.): H.v. Helmholtz, Philosophische Vorträge und Aufsätze, Akademie-Verlag, Berlin 1971, p. 301-335

Jahnke,H.N./Otte,M./Schminnes,B.: Introduction in: Jahnke/Otte (ed.): Epistemological and Social Problems of the Sciences in the Early Nineteenth Century, Dordrecht/London 1981, p. xi-xlii

Klein,F.: Über Arithmetisierung der Mathematik, Pädagogische Zeitung 1895

Mathews,J.: William Rowan Hamilton's Paper of 1837 on the Arithmetization of Analysis, Arch. Hist. ex. Sci. 19 (1978), p. 177-200

Schlesinger,L.: Über Gauss' Arbeiten zur Funktionentheorie, 1912, in: C.F.Gauss: Werke, Bd. X/2

Whitehead,A.N.: An Introduction to Mathematics, 12th Ed., London 1961

Williams,L.P.: Michael Faraday - A Biographie, London 1965

MATHEMATICS AND REVOLUTION FROM LACROIX TO CAUCHY

Luke Hodgkin

1. Some questions

This paper is the outline of a project arising out of more general work on mathematics in France in the period 1790-1830. The problem on which I shall focus can be stated quite simply. Cauchy's 1821 Cours d'analyse aimed explicitly at producing a 'revolution' in the language of analysis to supersede the erroneous earlier approach (which we could call 'classical'). The passage from Cauchy's introduction which is generally regarded as his manifesto reads:

> I have sought to give to the methods all the rigour which is demanded in geometry, in such a way as never to refer to reasons drawn from the generality of algebra... They tend to cause an indefinite validity to be attributed to algebraic formulae, while in reality the majority of these formulae hold only under certain conditions, and for certain values of the variables which they contain. By determining these conditions and values, and by fixing precisely the meaning of the notations I shall make use of, I shall dispel all uncertainty. (Cours d'analyse, p.ii)

The programme urged by Cauchy was widely adopted, and he is now, in the standard histories, credited with the systematic introduction of rigour into analysis. And yet some, but not all, of the older works survived. In particular, S.F.Lacroix's two main works on calculus (dating from 1795-1810) remained popular in France for a long time, several times revised and reprinted. Is this simply to be seen as conservatism - a delay in assimilating new ideas? Or are there some works which, after a revolution, have a greater survival value than others, insofar as there are alternative ways in which they can be read? [1)]

This limited question needs to be set in the context of two wider problems:

1. Internal. What significance do we give to the concept of a 'Cauchyan revolution' dated around 1821, given the points made by Grattan-Guinness (1970) about Cauchy's unoriginality (debt to Bolzano in particular), his failures in carrying out his programme, the persistence of infinitesimals, etc.?
2. External. An important transformation had already taken place in the practice of mathematics as a result of the Revolution - if by 'practice' we mean the ways in which the subject was studied, taught, communicated. Is there any link between this transformation and changes in the language and rules of mathematics - and if so, how do we describe it and account for it?

1) For the idea of different 'readings' of the same text, see the work of Pierre Macherey, in particular "An Interview with Pierre Macherey", Red Letters 5, London, n.d.

The second problem leads in turn to the methodological question repeatedly raised in this workshop of what is a legitimate programme for the social history of mathematics. Must it be restricted to mathematicians' 'external' life, to institutions and professions, applications and ideological content - 'what the internal history cannot explain', as some participants put it? Or can it go further and treat the way in which people do mathematics - in speech and writing - as part of their social discourse, interacting with other parts in ways which we try to understand? Is there, to take our example, a social analysis of the introduction of rigour in analysis which is not limited to describing a struggle between 'groups of supporters of competing paradigms' à la Kuhn, but which explains the meaning of rigour at different times for those who did or did not claim it? Can we extend a social explanation beyond the programmatic statements of mathematicians in the introductions to their books - often, as with Cauchy, statements of intention rather than accurate descriptions of the author's practice - and show what it meant at a given time to choose a more or less 'rigorous' method of finding the value of an infinite integral, or the Taylor series for a^x? I hope that in what follows I shall indicate the possibility of 'maximalist' answers to these questions - of an integrated description of mathematical practice within society.

2. Readability

To begin with, the question of what Cauchy's revolution may have done for subsequent reading and writing in analysis. It's an essential fact in the history of mathematics that much mathematical writing carries almost no explicit trace of its period. Lenin's 'Socialism equals soviets plus electrification' is clearly an equation which refers to a particular place and time; but, as we know, a textbook which poses the problem

$$x^2 + 10 = 7x$$

and finds the solution $x = 2$ or 5, can lay a trap for the historian in the apparent timelessness of its 'rightness', which in fact leaves many questions unanswered : Could the author have solved a wider class of equations - so that this one was chosen simply because it was elementary? How much wider? Would the readers of the text - at the time or later - have made the possible generalizations? I have mentioned these very simple questions just because we have the methods to deal with the history carefully in such a case. We don't always have them, because for some reason arguing from hindsight is more the norm once we get to something harder than quadratic equations. And yet every mathematical text is in some sense historically specific, written at a given time for a given audience. There are unwritten changes in the language of mathematics, or differences in understanding at a particular time, which elude us. Add to these notational changes (replacing 'a + b + c + ad inf.' by '$\sum_{i=0}^{\infty} a_i$') and the describable structural changes (changing use, and eventual disappearance of infinitesimals); we see any work being subjected over a period of time to a series of transformations in the context of reading which eventually lead to its becoming 'unreadable' in terms of a modern discourse and needing translation if it is to be used. (This has been done for example for some of the works of Riemann and S. Lie.)[2] But before that stage, any important shift changes the way in which what has previously been written can be read. Schwarz' work on distributions changed the way in which mathematicians read the earlier work of physicists on quantum mechanics, since there was now a reputable analytic discourse in which the delta function and its derivatives could be accommodated; previously such functions were either an acceptable piece of imagery which could be rigourously justified if one wanted to, or (for von Neumann) dangerous nonsense.

We need a language, a theory for describing these changes in the discourse of mathematics, in the rules for writing and reading. Some are abrupt and overt, some slow and almost unconscious. 'f|A is continuous' has come to mean something different from its predecessor 'f is continuous on A', with the spread of categorical ideas. A reader will bring a different set of assumptions to bear on such a statement, so that although its 'logical' meaning may be seen as remaining constant,

2) Riemann's '<u>Über die Hypothesen</u>...' is edited and commented (among other works) in M. Spivak's <u>Differential Geometry</u>; some of S. Lie's main papers have been edited with commentaries by R. Hermann. In both cases it's a question of updating texts for working mathematicians, though there are interesting differences of approach.
The <u>general</u> question of when and how a work 'drops out of circulation' is a fascinating one which would repay particular study.

its meaning in terms of the mathematician's practice has changed.

The specific point about Cauchy's work which transcends the limitations raised above, is that it made definitive one of these shifts in the domain of what is readable and how it is read. It's better to use such relativistic language here than to talk of 'rigour', whose meaning can vary widely. Certainly the Cours d'analyse (with Cauchy's associated writings) do not carry the whole responsibility for the shift; we must take in the previous work of Gauss, of Bolzano and others, and the subsequent tidying up of Cauchy's failures on uniform convergence and elsewhere. But they have the particular status of propaganda works which announce to the students who learn from them (explicitly or implicitly): these are the rules for forming correct statements, disregard anything which seems to be constructed otherwise. We have evidence from Abel and others that there were students who accepted the message completely. A modern parallel is the work of Bourbaki - which, significantly, does not need to advise the student to reject non-Bourbakist work, since it creates a universal and cohesive system of practice which it's difficult for a student to get out of, once inside.

If we now ask what were the new rules on how texts should be read instituted by Cauchy, one in particular has been clarified by Grattan-Guinness: the requirement that the expression for the sum of a series should mean its sum by 'orthodox' or 'regular' summation (i.e. without rearranging terms), and that such a sum should be convergent. Also, series of functions should carry with them, as part of their description, the limits within which they converged. (Again, even for those of Cauchy's contemporaries who didn't accept this requirement, its presence, for acceptance or rejection, constituted a new point of reference.) But this is only one element among others, which need to be made explicit: on continuity, the use of infinitesimals, and so on. The whole structure, when reconstructed, will involve many such elements and linkages.

3. Revolutions

There is an obvious answer to the 'external' question (the influence of the revolutions in French society), which is part of the historical tradition. The first ten years of the Revolution created teaching institutions, notably the Écoles Centrales and the Polytechnique, in which mathematics had a new importance for a much wider class of students, the cadres of the new army and civil service.[3] Immediately the teaching of analysis became a problem area. (The connection

3) This needs to be qualified, since there were certainly some changes in curriculum in the last years of the Ancien Regime at the colleges (probably limited), and more importantly at the military schools. Also, the class of students seeking education seems to have been changing; see Mornet for these points. The reform of the year III did however mark a radical break in institutionalizing the links between school and state (especially the army) and in the importance which it gave to mathematics in particular.

with teaching is the significant element here since, as is well known, the foundations of calculus - 'metaphysique du calcul' - had been a problem since its origin.) Lagrange's prestigious course at the Polytechnique, which was based on his ideas of how analysis should be made problem-free for students, had to be supplemented by a simpler one which explained what he was doing. Over the next few years a number of textbooks, in particular Lagrange's Leçons sur le calcul des fonctions (1801-6) and Lacroix's Traité élémentaire du calcul (1802) and Traité du calcul (1797-1800), set out to make classical analysis sounder and more accessible, aims which were not always harmonized. Cauchy's work is in this tradition; devised for the post-restoration Polytechnique - which, like many of the institutions of the Restoration, shared more with its Napoleonic predecessor than with the old Bourbon system - it succeeded where the earlier works had failed.

This account does place the events in the context of the changed institutions of post-revolutionary France. But it has the limitations of much 'external' history of mathematics which I outlined earlier; it fails to explain why the change in analytic discourse took the form it did - why 'limit-avoidance' and similar methods were chosen for dealing with the problems. Such questions are open to the standard criticisms of 'why' questions in history - the whole collection of problems associated with historical causation and determinism. And yet any description which does not attempt to explain in some terms is historically sterile. One fresh approach, which relates particularly well to the kinds of problems I have raised, is provided by the idea of 'archaeology' which Michel Foucault has outlined in The Archaeology of Knowledge, and applied to various fields in history of science - in our period in particular - in The Birth of the Clinic and The Order of Things.

It is almost impossible to summarize Foucault's method in a short space; but one central feature is the attempt to look not for 'traditions' linking work done in a science at different times, but for a 'law of dispersion' which characterizes the (contradictory) state of a science at one particular time, the fundamental units of study being all statements made within or about the science (the 'archive'):

> Whenever one can describe, between a number of statements, such a system of dispersion, whenever, between objects, types of statement, concepts, or thematic choices, one can define a regularity (an order, correlations, positions and functionings, transformations), we will say, for the sake of convenience, that we are dealing with a discursive formation - thus avoiding words that are already overladen with conditions and consequences, and in any case inadequate to the task of designating such a dispersion, such as 'science', 'ideology', 'theory', or 'domain of objectivity'. The conditions to which the elements of this division (objects, mode of statement, concepts, thematic choices) are subjected we shall call the rules of formation. The rules of formation are conditions of existence (but also of coexistence, maintenance, modification, and disappearance) in a given discursive division. (Archaeology of Knowledge, p.38)

The aim as applied here would then be to draw on the largest possible field of statements available to characterize two discursive formations - the 'classical' analysis of the 1780's and its successor, a 'postclassical' analysis whose boundaries still need to be determined. (Lagrange after 1795 is on this model no longer writing within 'classical' analysis; indeed it is important that statements by the same author at different times, or in different contexts, may not belong to the same discursive formation.) The statements available should extend beyond research papers and advanced textbooks to elementary textbooks, teaching manuals, examination questions (and answers, where we have them), and more generally to everything which bears on the way in which analysis was practised and spoken about during the period. Examples which come to mind are the educational projects of Condorcet, Lepeletier and others, the actual educational decrees of the Convention and Directory, and what we know of their application (records of the Écoles Centrales in particular),[4] the accounts of progress at the Polytechnique, meetings of learned and not so learned societies. Further, given that many figures whose life was far from mathematical (military men, politicians, etc.) passed at this period through an educational process in which mathematics had a decided importance, we may recover quite specific attitudes to mathematics in memoirs of the period

4) One question which occurs is precisely how much mathematics was taught in the Écoles Centrales, and at what level. By the formal programme of the law of 3 Brumaire IV it should have been confined to 1/3 of the teaching in the second year (out of three); and in Lacroix's École des Quatre-Nations this seems to have happened (Crosland, 1969). But in many provincial schools some at least of the teachers were missing, so the programme would differ from the prescribed one. At Grenoble, by Stendhal's account, there seems to have been much more mathematics.

(see below, §4 for an example). Further, (and here we return to the original question of this section), the breadth of the material may make possible the construction of the important linkage which Foucault avoids[5] - the relation between the change in the discursive formation and the political and ideological changes in French society.

Some tentative thoughts in this direction are:

1. It would appear that in the 18th century the space of mathematical <u>objects</u> has certain boundaries - about which there may be dispute. (A curve traced at random on a sheet of paper is not a curve in the sense of mathematics, wrote d'Alembert in the <u>Encyclopedie</u> (article <u>Courbe</u>); others thought differently.) But the space of <u>operations</u> on these objects, once they have been so restricted, is free; so functions exchange with Taylor series, and surfaces <u>are</u> also polyhedra with infinitely small faces. The 19th century seeks to extend the rules for entry into the domain of mathematical objects (discontinuous or many-valued functions, for example), while the criteria for entry become more technical and less 'philosophical'. At the same time it moves to a position where within this domain each operation has a restricted subdomain

5) D. Lecourt (1972) has criticized Foucault specifically from a Marxist viewpoint for avoiding such a linkage, while claiming an importance for Foucault's work within his own Marxist perspective.

of validity, and is the result of labour rather than free exchange. Even over the mathematician's work, Adam Smith casts a shadow.[6)]

2. It is fairly clear from the writings of political thinkers some of whom (like d'Alembert and Condorcet) were themselves mathematicians, that at the outset of the Revolution mathematics had a specially privileged position among sciences as the model of 'rationality' and the opponent of obscurantism. This position, which seems to belong particularly to the 18th century, lost both influence and credibility around 1800 with the growing importance of chemistry in particular, and the fall of the political tendency which the Ideologues represented; by 1815, mathematics was one important science among others, without a particular moral weight. It therefore needed guarantees of a new kind, not in terms of abstract rationality, but of unproblematical technique.

3. It would be fascinating, in terms of the 'sociology of knowledge' programme, to juxtapose these tendencies with the very heterogeneous reaction against the Enlightenment which is generally called Romantic, and which often involved a denial of the 'open space' which the Enlightenment took for granted. Blake's attitude to Newton is well-known; equally

6) Explicitly, of course, in the classic example of Prony's application of the division of labour to the calculation of tables.

suggestive is Saint-Simon:

> Great men of all the ages, Newton and Leibniz, Voltaire and Rousseau, do you know what you are great in? You are great in blindness... for having thought that civilization was the social destiny of the human race.[7)]

Did Cauchy's restrictions draw their strength from an increasing belief in the impossibility of freedom - reinforced by his peculiar political convictions?

4. An example

Among the 'statements' we might examine for evidence on the practice of mathematics in the 1790's, Stendhal's autobiographical fragment The Life of Henry Brulard is exemplary. At a very far remove from the world of research papers, Stendhal describes what it was like to study mathematics in Grenoble in the 1790's - indeed to have a 'love-affair' with the subject. Importantly, mathematics represented (as mentioned before) republicanism, rationality; also the means of escape from Grenoble to Paris and the Polytechnique. Added to these we have the real presences of teachers, textbooks, and difficulties with the subject, which Stendhal never dissolves into a fraudulent unity. The collection of 'statements' which are held together in a passage such as the following is there-

7) Quoted by L. Fèbvre in his article on 'Civilization' (Fèbvre 1973, p.239). The reactionary nature of early French romanticism (e.g. Chateaubriand) should be taken into account.

fore exceptionally rich:

> I loved mathematics all the more because of my increased contempt for my teachers, MM. Dupuy and Chabert. In spite of the grandiloquence and urbanity, the suave and dignified air that M. Dupuy assumed when he spoke to anyone, I had enough shrewdness to guess that he was infinitely more of an ignoramus than M. Chabert. M. Chabert, who in the social hierarchy of the bourgeoisie of Grenoble stood so far below M. Dupuy, sometimes on a Sunday or Thursday morning would take a volume of Euler or ... and resolutely tackle difficulties ....
>
> My enthusiasm for mathematics may have had as its principal basis my loathing for hypocrisy, which for me meant my aunt Séraphie, Mme Vignon and their priests.
>
> In my view, hypocrisy was impossible in mathematics and, in my youthful simplicity, I thought it must be so in all the sciences to which, as I had been told, they were applied. What a shock for me to discover that nobody could explain to me how it happened that: minus multiplied by minus equals plus ($- \times - = +$)! (This is one of the fundamental bases for the science known as algebra.)
>
> Not only did people not explain this difficulty to me (and it is surely explainable, since it leads to truth), but, what was much worse, they explained it on grounds which were evidently far from clear to themselves.
>
> M. Chabert, when I pressed him, grew confused, repeating his lesson, that very lesson against which I had raised objections, and eventually seemed to tell me: 'But it's the custom; everybody accepts this explanation. Why, Euler and Lagrange, who presumably were as good as you are, accepted it!' (From The Life of Henry Brulard (1973), 299-302.)

Mathematics indeed, in becoming a study, not for 'the masses' but for a relatively large and unformed élite, had to abandon its position of bastion against 'hypocrisy' and learn, like the priests, how to get around contradictions and difficulties. The 'jacobin' feeling that 'it is surely explainable, since it leads to truth' had to be dealt with, better suppressed, not by the crude methods of Dupuy and Chabert, but by taming the role of the imagination. And the great textbooks of the period had a crucial influence on this process.

5. Lacroix

This leads us to the very interesting historical figure of Lacroix, a participant in the institutions and thought of the whole period. In the field of education, where his views are best expressed by the Essai sur l'enseignement en général et sur celui des mathèmatiques en particulier (1805), he had a great deal of experience, of influence, and of commitment to 'revolutionary' ideas and methods. His equally important mathematical writing has been misread by a history which stresses clear innovation in content, ignoring style and discourse. And yet if we compare them with the more original works on which he - very conscientiously, naming his sources - draws (Clairaut, Euler, Lagrange) as well as those which he definitely avoids (Bézout), the sense of limitation, the common sense of the 19th century, is already there in embryo, contrasted with the assured rationalism of his 18th century predecessors. The influences on Lacroix were diverse; his much more talented (from the research standpoint) contemporaries Lagrange, Legendre and in particular Laplace, to whose work the Traité du calcul is meant to be an introduction; previous educational thinkers particularly Condorcet and Rousseau; and the experience of a teacher and educational 'politician', mostly at the École Centrale des Quatre-Nations, which seems to have been a showpiece for the system of Écoles Centrales. These combine to produce a mathematical discourse in which many contradictory elements coexist. For example, he more than once states the very 'classical' view that the differentiability of functions is a fact 'antérieur a toute hypothèse' like the falling of heavy bodies to the earth. On the other hand in his textbooks he practises limit-avoiding arguments, without including them in general programme such as Cauchy was to introduce. (He also avoids infinitesimals, which Cauchy did not.) I think that the unity of these elements lies in a pragmatic belief in the correctness of doing mathematics, simply and without philosophical underpinnings. If we are not too fanciful (as say Euler was) about the calculations we do, we avoid the dangerous recourse to the outside for justification. It is interesting to see Lacroix quoting with approval (in the Essai sur l'enseignement) the following passage from Saurin (1725):

> Philosophers and those who principally study the higher sciences honour Geometry when they deign to apply themselves to it; but, full of confidence in their enlightenment, they wish at first to clarify everything, as if everything was left obscure. With the greatest enlightenment and the best intentions, they might spoil everything in giving too much weight , not to reason but to the reasoning (non à la raison, mais aux raisonnemens)... Our calculations have not so much need of clarification as one thinks; they carry their own light with them; and it is normally from within them (de leur sein même) that issues forth all the light that one can throw on them, and that the subject being treated can receive ... It is never calculation which deceives us when it is well done; it does not need to be supported by reasonings; but, normally, it is reasonings which deceive us, and which must not decide us except insofar as they are supported by calculation.

Saurin's text, which belongs to an 18th century discussion on the guarantees of mathematical reasoning, has become transformed by being placed in a discourse on education eighty years later. A key sentence is: 'It is never calculation which deceives us when it is well done'. In the perspective of 1725, 'when it is well done' is a routine point of qualification;

by 1805, with the experience of repeated controversies in analysis about what are 'well done calculations', and of problems in conveying their value to the student (e.g. Stendhal) who is the main focus of attention for Lacroix, that qualification has become a _prise de position_ closely related to that of Cauchy. 'Reasonings' must give way to 'calculations', indeed, but not everything which has the outward appearance of a calculation can be accepted. In these traces, I think, we can find the first signs of the new discursive formation.

The real test lies in the mathematical texts, and it is instructive to consider Lacroix's derivation of the Taylor series for a^x in relation to those of Euler (in the _Introductio ad analysin_) and Cauchy. (Ideally one should do this at length and include a wide range of other writers, but space forbids this here.) The three are logically very close; all of them use arguments which would be faulted from a modern standpoint. But there is an attempt by Lacroix and Cauchy to ensure a restriction for the domain of validity of operations whose staggering absence is one of the impressive features (in a sense) of Euler's proof. To take one example: Euler uses an infinitesimal called 'ω' in his proof, and then as an example sets $\omega = 1/1000000$. This fluidity has vanished by the time of Lacroix.

Lacroix's proof (from the _Elementary Treatise_ in Babbage and Herschel's translation, p.24), reads:

> 23. Those functions which are not comprehended in the enumeration made in No. 14, are called transcendents. The exponential function $u = a^x$ is the most simple of this sort. When we substitute $x + dx$ instead of x, the differ-

ence becomes

$$a^{x+dx} - a^x = a^x(a^{dx} - 1);$$

and in order to express it according to the powers of dx, we make a = 1 + b, when it becomes

$$a^{dx} = (1 + b)^{dx} = 1 + \frac{dx}{1} b + \frac{dx(dx - 1)}{1.2} b^2 + \frac{dx(dx - 1)(dx - 2)}{1.2.3} b^3 + \&c.$$

Whence

$$a^{dx} - 1 = \left\{\frac{dx}{1} b + \frac{dx(dx-1)}{1.2} b^2 + \frac{dx(dx-1)(dx-2)}{1.2.3} b^3 + \&c.\right\}$$

and arranging this according to the powers of dx,

$$a^{dx} - 1 = dx\left(\frac{b}{1} - \frac{b^2}{2} + \frac{b^3}{3} - \&c\right) + \&c.$$

replacing b by its value a - 1, there results (5)

$$d.a^x = a^x dx \left(\frac{a-1}{1} - \frac{(a-1)^2}{2} + \frac{(a-1)^3}{3} - \&c.\right);$$

and making

$$k = \frac{a-1}{1} - \frac{(a-1)^2}{2} + \frac{(a-1)^3}{3} - \&c.$$

we have $d.a^x = k\ a^x\ dx$.

This is the form of the differential of the proposed function, and we shall soon find a new expression for the constant quantity k.

Lacroix deduces from this the Taylor series for a^x, defines e, and shows that $e^k = a$. This gives the promised expression for $k = \log.a/\log.e$. The derivation above is conventional. enough. The notation does hide some points. For example, dx is not an infinitesimal - this is clear from the beginning of the treatise. The final formula expresses $d.a^x$ as a function of two independent finite variables x and dx (an approach common in modern differential geometry). Hence expressions like a^{dx} are not dependent on the theory of infinitesimals as they are for some other writers of the period.

The derivative of the function u is found by taking u' (the value of u at x + dx), expanding it in powers of dx, and isolating the term of first degree in dx, which is precisely du.

This is equivalent to the limit definition for du/dx, because for Lacroix all functions can be expanded in power series.

A problem is that the various series (e.g. for k = log.a/log.e) are written down with no statements about their convergence. It is clear from other places in the book - as usual at this period - that this is not because it's not known that divergent series exist, are a nuisance, and should be avoided. However, this is definitely an example of the kind of unlimited formalistic statement which Cauchy wished to banish. The difference between Lacroix and Euler is that Euler, at the comparable point in the <u>Introductio ad analysin</u>, takes the example a = 10 and claims that, in some sense to be made clear later,

$$\log.10 = \frac{9}{1} - \frac{9^2}{2} + \frac{9^3}{3} - \&c.$$

The <u>absence</u> of such a statement can be taken as an implicit instruction from Lacroix to the reader not to specialize the formula in this way.

To summarize: more evidence is needed, of course, before the discursive formations involved can be analysed, distinguished, and related to the practice of mathematics in teaching and elsewhere. For example, any account of the period which does not mention descriptive geometry, the 'ruling subject' in the Polytechnique, if only to justify its omission, is very incomplete. With this qualification, we can give a quite specific importance to Cauchy's role in recasting the language of analysis; while at the same time fixing the major transition from classical to postclassical in the years 1795 -

1800 when the conditions of practice of mathematics became transformed so that some linguistic reform became essential. Because of his closeness to those changes in practice, Lacroix's language was new enough to look forward to future readings; and hence his work was able to last.

MAIN BOOKS CONSULTED

Textbooks

Cauchy, A.L., (1821) Cours d'analyse de l'École Royale Polytechnique, Paris.

____________ (1823) Résumé des leçons données a l'École Royale Polytechnique sur le calcul infinitésimal, Paris.

Clairaut, A.C. (1749) Élémens d'algèbre, Paris.

Euler, L. (1748) Introductio in analysin infinitorum, Lausanne; also French translation (1796-7) 'avec notes et éclaircissements' by J.B. Labey, Paris.

________ (1755) Institutiones calculi differentialis cum ejus usu in Analysi Finitorum ac Doctrina Serierum, St. Petersburg/ Berlin.

Lacroix, S.F. (1797-1800) Traité du Calcul différentiel et du Calcul intégral, Paris.

____________ (1802) Traité élémentaire du Calcul différentiel et du Calcul intégral, Paris. English translation by Babbage, Peacock and Herschel, (1860), Cambridge.

Lagrange, J.L. (1804) Leçons sur le calcul des fonctions, Paris.

Others

Balibar, Renée (1974) Les Français Fictifs: le rapport du style national au français national, Paris.

Clagett, Marshall (1969) ed., Critical Problems in the History of Science (pp. 291-320), Madison.

Crosland, Maurice (1967) The Society of Arcueil, London.

________ (1969) ed., Science in France in the Revolutionary Era, described by Thomas Bugge, Cambridge, Mass. and London.

École Polytechnique (1895) Livre du Centenaire, 1794-1894, t.l. (L'École et la Science), Paris.

Euler, L. (1787-1789) Lettres à une Princesse Allemande, nouvelle édition avec des Additions par. MM. le Marquis de Condorcet et Lacroix, Paris.

Fèbvre, L. (1973) A new kind of history (ed. Peter Burke), London.

de la Fontainerie, F. (1932) tr. and ed., French Liberalism and Education in the Eighteenth Century. The Writings of La Chalotais, Turgot, Diderot and Condorcet on National Education, New York and London.

Foucault, M. (1970) The Order of Things, London.

________ (1972) The Archaeology of Knowledge, London.

________ (1973) The Birth of the Clinic, London.

Fourcy, A. (1828) Histoire de l'École Polytechnique, Paris.

Gramsci, A. (1971) The Formation of the Intellectuals, in Selections from the Prison Notebooks, tr. Q.Hoare and G.Nowell-Smith, London.

Grattan-Guinness, I. (1970) The development of the foundations of mathematical analysis from Euler to Riemann, Cambridge Mass. and London.

Guiomar, J.-Y. (1974) L'idéologie nationale, Ed. Champ Libre.

Hippeau, C. (1883) *L'Instruction Publique en France pendant la Révolution: Débats Legislatifs*, Paris.

Israel, G. and Negrini, P. (1973) "La Rivoluzione Francese e la Scienza", *Scientia*.

Lacroix, S.F. (1805) *Essai sur l'enseignement en général et sur celui des mathématiques en particulier*, Paris.

Lakatos, I. (1978) *Mathematics, Science, Epistemology*, Cambridge.

Lecourt, D. (1972) *Pour une critique de l'épistémologie: Bachelard, Canguilhem, Foucault*, Paris.

Liard, L. (1888) *L'Enseignement supérieur en France, 1789-1889*, Paris.

Mornet, D. (1933) *Les origines intellectuelles de la Révolution française*, Paris.

Stendhal (1973) *The Life of Henry Brulard*, tr. Jean Stewart and B.C.J.G. Knight, Harmondsworth.

Taton, R. (1959) "Condorcet et Sylvestre-François Lacroix", *Rev.Hist.Sci*.12, 128-158, 243-262.

PART II

THE PROFESSIONALIZATION OF MATHEMATICS AND ITS EDUCATIONAL CONTEXT

INTRODUCTION

Ivo Schneider

When one speaks publicly today of a "pro," it is almost always in association with sports. The professional athlete is distinguished from the amateur, the most important difference between them lying in the payment of the professional for his performance. The prerequisite for this payment is of course the pro's fairly high performance level, whose attainment and maintenance demands a training program that absorbs a large part of the time normally available for one's occupation.

Quite naturally some historians of science have attempted to associate the idea of a sports "pro" with the concept of professionalization, speaking of it in the modern sociological sense as a process in the development of science. More precisely, professionalization has to do with a transition phase in the development of mathematics and science, at whose end it was possible to pursue science for its own sake. This transition phase has generally been placed in the 19th century, its beginning and duration being set at different times in different countries depending on local political and social conditions. Thus in Germany one finds the beginning of professionalization in the first half of the 19th century, together with the establishment of a new educational system that culminated in the new Prussian universities, while in the United States it is placed in the 1860's,[1] reflecting the delayed scientific development of that country vis-à-vis Europe.

It is tempting to understand this process as the replacement of the amateurs, who had until then been responsible for the progress of science, by a new group of professional scientists. In this way one looks for criteria that will characterize or distinguish this new professional group. It is clear, however, that there is no single set of criteria for characterizing a professional scientist that will be suitable for all scientific disciplines. One reason for this impossibility is that there are not only various degrees, but also various forms of professionalization, depending on the subject and on the embedding of the educational system within the political and social system of a particular nation. Even more important, at the end of the process of professionalization one does not find, as in sports, two groups facing each other that one can call amateurs and professionals; rather, the professional scientist stands alone. This phenomenon is central to the interpretation of a group of sociologists of science whose principal representative I will take as Magali Sarfatti Larson[2]. Larson sees "professionalization as the process by which producers of special services sought to constitute and control a market for their expertise" and "also as a collective assertion of special social status and as a collective process of upward social mobility"[3].

In this sense, Larson says, "the professionalization movements of the nineteenth century prefigure the general restructuring of social inequality in contemporary capitalist societies: the 'backbone' is the occupational hierarchy, that is, a differential system of competences and rewards; the central principle of legitimacy is founded on the achievement of socially recognized expertise, or, more simply, on a system of education and credentialing. Professionalization is thus an attempt to translate one order of scarce resources - special knowledge and skills - into another - social and economic rewards. To maintain scarcity implies a tendency to monopoly: monopoly of expertise in the market, monopoly of status in a system of stratification"[4]. In particular the intended monopoly of status "accentuates the role that educational systems play in different structures of social inequality".

Different degrees of professionalization are arranged by Larson according to the degree of their monopolization of the market, to attaining social status and work autonomy which seems to him to reach its highest point in medicine. Two structural elements are necessary for this: "a specific body of knowledge, including techniques and skills", and "a market of services". Both elements vary nationally and historically as well as according to the specific profession[5].

The essence of the professionalization process, in Larson's view, is the training of a professional producer, which is indissolubly tied to the development of the universities. The result of this is the "monopolization of competence and the demonstration that this competence is superior to others." One of the most important conditions for the legitimation of this training monopoly is that it permits the training of anyone who seeks it and is able to produce the necessary work required for it. This "meritocratic legitimation" process becomes effective only after the establishment of a bourgeois hegemony, which permits an "open, although hierarchical system of education."[6]

Larson's concept of professionalization appears to be especially useful, because it avoids the difficulties that result from the common tendency to evaluate various national situations in terms of a particularly successful model of professionalization. On such a basis, for example, one would assign a hopeless backwardness to England in professionalization in the 19th century in spite of its obvious proficiency in fields like astronomy and biology.

One of the most important aspects of the professionalization process is research, which Ben-David has called perhaps the highest product generated by the professional scientist. Proceeding from the role of research in the professionalization process, one next asks how research and forms of professional science that can serve as a model were operative before the beginning of the professionalization process. Of course, there were mathematical research and forms of professional activity in mathematics in earlier times that influenced positively or negatively the structure of the forms that developed in the nineteenth century. This is especially to be expected in those places where the educational system was taken over relatively unchanged from the 18th to the 19th century. In this context the first article, "Forms of Professional Activity in Mathematics before the 19th century," seeks to develop such models and to describe the corresponding objectives of mathematics in the two centuries preceding the 19th.

In the interval between 1600 and 1800 the dominant mode of occupation in mathematics was based on the model of the artisan, working within the structure of a guild. Mathematics remained a static subject, the content of which was organized in recipe-like methods applied to a canon of nearly standardized exercises. The increased influx of Greek mathematics in the 16th century and the subsequent attempts to reconstruct the missing parts of Greek mathematics opened up new areas of mathematical activity. Beginning in the second half of the 17th century, first the private tutor (following the model of the artisan) and then the academician developed as forms of professional activity suitable for these areas. Of these two forms, which continued up to the 19th century, the private tutors failed as a professional form because the attempt to transform the results of the new mathematics into a subject taught in a recipe-like manner caused an ever widening division between teaching and research which could be bridged only by autodidactical study. In addition the market for the product "mathematics" was still much too small to allow more than a very few to work in it. Connected with this is the comparatively low prestige and social status accorded to mathematics which caused serious problems in recruiting new candidates for the academies.

In order to overcome these obstacles to the establishment of a professional mathematics, one had to create a new social image of the subject and with it a considerably enlarged market. The second and following articles are mainly concerned to show how in the different European countries with their specific economic and political conditions, as well as intellectual climates, such an enlarged market for mathematics was constituted and to what degree it could be controled. This is to say that we observe different stategies used by the rising bourgeoisie to gain influence over the educational system as the most important means to develop a new market for mathematics and by it to professionalize mathematics. Gert Schubring's "conception of pure mathematics as an instrument in the professionalization of mathematics" treats this process for the first decades of the 19th century in Germany. The bourgeoisie in Prussia used a neohumanistic ideology in order to reform the entire educational system which was itself instrumental in replacing a mercantile by a bourgeois middle-class production system.

The aim of upward social mobility within an intended hierarchy of occupations including scientific professions was to be secured by proofs of competence, that is by meritocratic legitimation. The ability to do research was already considered by 18th century academicians as the most important criterium for competence in science. Thus the neohumanistic ideology stressed the importance of research by instituting a "research imperative"[7] for everybody involved in the system of higher education as a teacher. The fact, that the bourgeoisie succeeded in making this the Prussian state ideology and that the Prussian state required complete control of the educational system, paved the way for the professionalization of science in Prussia and Germany. The Prussian state monopoly of the educational system automatically secured a monopoly in expertise and status for the new scientific professions. Due to a prevailing neo-Kantian, idealistic tradition which opened a market for pure mathematics within the educational market, mathematics became one of the subjects that figured prominently in this educational system; it was freed from the necessity to justify itself by its applicability and utility to other domains.

It is regrettable that we cannot include a special study of the professionalization process in France. Schubring, however compares France with Germany and, utilizing an understanding of professionalization different from Sarfatti Larson's, concludes that in France there is no professional mathematics before 1870[8]. A detailed account of the French situation would have to consider the fact that French mathematicians produced more than 80% of the mathematics published in the first three or four decades of the 19th century[9], that in spite of a climate favourable only for applied mathematics much of the

French mathematics labelled "applied" would have been considered as pure mathematics by German mathematicians, and that most of the leading French mathematicians enjoyed a very high social status. In other words the fact that the international "market" of mathematics was dominated by French mathematical production at least up to 1840, hints at the likelihood of a process of professionalization in French mathematics before 1870. This still unravelled process must have followed a pattern different from that in Germany. This is recognizable from the stress on applied mathematics, the prevailing ideology of spiritualism, and the comparatively low degree of institutionalization in French mathematical research. Recent articles by Crosland, Fox, and Shinn[10] offer at least some material in order to clarify this process.

The situation in 19th-century England differs from that in both Germany and France. The English educational system was composed of a large number of different independent private schools and two universities, to most of which access was controled according to one's social background and/or qualifications. Considering the relatively high stability of this system the possibilities of creating a new market for mathematics and by it of professionalizing mathematics were restricted, the only means being the foundation of new private schools or universities or a modest reform in the curricula of the old ones. In the light of these possibilities, neither a statement like, "It is wrong to conclude that science achieved <u>full</u> professional status in 19th century Britain,"[11] nor the professional performance of men like John Herschel, George Boole or Charles Darwin, are astounding. What could be achieved in the English situation was a restricted professionalization within a part of the educational system. This is dealt with for the domain outside the universities by Leo Rogers. Rogers maintains that, on the one hand professional forms of mathematical activity like the private tutor and the mathematical practitioner continued and, on the other hand, that a climate created by a rationalist, materialist philosophy, and by the necessities of the Industrial Revolution caused a change in the social image of science and mathematics. As a result the rising English middle class, especially those from a non-conformist background, who were not allowed to study at Oxford or Cambridge, established new institutions with special emphasis on the teaching of science and mathematics. Thus we observe an enlarged market for an essentially practical, applied mathematics which was transmitted to a wider public by a new class of teachers. These teachers present a professional group of mathematicians different from the German Gymnasia-teachers. There was no special "research imperative" for the English mathematics teachers. However, they backed up their claim to expertise and competence by shaping a new methodology for disseminating mathematical knowledge. Many of these mathematics teachers took part in the later development of a national system of universal education and so prepared the way for a higher degree of monopolization and professionalization in mathematics.

Phil Enros shows, taking Cambridge as his example, how, on the higher level of the English universities, a social demand for mathematical research was developed. According to the prevailing ideology of Cambridge at the beginning of the 19th century, mathematics was considered an essential subject for educating the future leaders of the Commonwealth. This close relationship between mathematics and leadership was extended to research by Cambridge students, who perhaps were impressed by the successes of Napoleonic France. Contact with the most recent research mathematics would improve the quality of leadership. As a first step, the students took care to get better information about Continental and especially French analytics. As a second step, the Cambridge tutoring and examination system in mathematics was adapted to the more advanced French level. Even if Cambridge did not get interested in mathematical research for its own sake until much later, its privileged position in the English educational system sufficed to stimulate a new concern for mathematical research in England which finally led to the beginnings of a professional mathematics in England.

Another factor influencing the professionalization process in mathematics is that once it has started in a particular country, that is that mathematics is understood as a socially relevant product, the mathematical production of this country must compete with that of other countries within its own national market. This competitiveness, together with national prestige, makes the "market leader" a good model for other countries which are behind in the professionalization process. This can be seen especially in Umberto Bottazini's "Mathematics in a unified Italy." In the middle of the 19th century, Germany had taken the lead in mathematics from France. The German model of professionalization in mathematics seemed the more acceptable to the Italians because the political and social developments in Italy and Germany showed considerable similarities, most noticeably in the unification movements in both countries. Thus the typical German combination of teaching and research in mathematics in universities and secondary schools was adopted by Italian mathematicians, who had abandoned the prevailing French tradition in applied mathematics, and who had considerable influence in shaping the new Italian educational system and with it their own market.

The concluding article, Horst Eckart Gross' "The employment of mathematicians in insurance companies in the 19th century," considers a more advanced level of the professionalization process in mathematics which extends the educational framework. When the foundation of insurance companies in 19th-century Germany opened a new market for some kind of expertise, it was by no means clear who would have access to this market, since it demands the solution of some comparatively simple mathematical problems with a host of nonmathematical problems. The formation of courses in insurance mathematics at the university level indicates that professional mathematicians were attempting to conquer and monopolize this new market for themselves.

Gross' analysis of the working process of actuaries in the 19th century can be seen as a case study of the working process of the mathematician outside the educational system. This may be applicable to especially industry in the 20th century, where, similar to the insurance scene, mathematical expertise is used to solve a combination of mathematical and nonmathematical problems. Of course, this development demands a new type of mathematics and new forms of research in this mathematics, which can claim a new dimension of social relevance.

Notes

1. See Nathan Reingold, Definitions and Speculations: The Professionalization of Science in America in the Nineteenth Century, in: The Pursuit of Knowledge in the Early American Republic (eds. A. Oleson and S. C. Brown), Baltimore - London 1976, pp. 33-69.

2. Magali Sarfatti Larson, The Rise of Professionalism, Berkeley - Los Angeles - London 1977.

3. Larson, op. cit., p. XVI.

4. Larson, op. cit., p. XVII.

5. Larson, op. cit., p. 49 f.

6. Larson, op. cit., p. 51.

7. See R. Stephen Turner, "The Growth of Professorial Research in Prussia, 1818 to 1848 - Causes and Context," Historical Studies in the Physical Sciences, 3 (1971), pp. 137-182.

8. Compare Gert Schubring, "Bedingungen der Professionalisierung von Wissenschaft. Eine vergleichende Übersicht zu Frankreich und Preußen", Lendemains, no. 19 (1980), pp. 125-135.

9. See e.g. Ivor Grattan Guiness, "Mathematical Physics in France, 1800 - 1835", in: Epistemological and Social Problems of the Sciences in the Early Nineteenth Century (eds. H. N. Jahnke and M. Otte), Dordrecht 1981, pp. 349 - 370.

10. See Maurice Crosland, "The Development of a Professional Career in Science in France," in: The Emergence of Science in Western Europe (ed. M. Crosland), London 1975, pp. 139-159; Robert Fox, "Scientific Enterprise and the Patronage of Research in France 1800 - 1870", Minerva, 11 (1973), pp. 442-473; and Terry Shinn, "The French Science Faculty System, 1808-1914: Institutional Change and Research Potential in Mathematics and the Physical Sciences," Historical Studies in the Physical Sciences, 10 (1979), pp. 271-332.

11. Victorian Science (eds. G. Basalla, W. Coleman, and R. H. Kargon), N.Y. 1970, Introduction p. 9.

FORMS OF PROFESSIONAL ACTIVITY IN MATHEMATICS BEFORE THE NINETEENTH CENTURY

Ivo Schneider

In 1977 a questionnaire was sent to all mathematics graduates of the BRD asking what their current professional activity was. This very fact demonstrates that while the existence of a specialized training program forms the basis for recognizing the professional status of the mathematician in current society, the manner and extent of the ever-changing and expanding possibilities for utilizing the special skills of this professional group must be constantly reexamined. These shifts are even recognizable in the instructional program itself, for the course offerings in the various institutions of learning are closely correlated with current research, which influences not only the composition and selection of materials for required courses of study but also the introduction of new disciplines. That even areas considered as the core of the educational curriculum are subject to gradual change is indicated by the progressively diminishing course offerings in number theory and geometry, which were once considered classical areas of study, but are scarcely even regarded in current research. Any attempt to delimit the current professional structure of mathematics must take into consideration this dynamic change in the content of educational programs and employment opportunities.

On the basis of that "backslapping" fraternal relation between mathematicians and the history of their subject, which, individual details notwithstanding, permits figures such as Archimedes, Newton, Euler, Gauß or Cauchy to be accepted as permanent members of the club of respectable mathematicians, it is assumed as obvious, even unquestionable, that the factors conditioning the actual conduct of mathematical research and the ability to engage in mathematics are nearly the same at all times; and accordingly it is not regarded as necessary to inquire further into these conditions. In point of fact, however, the preconditions for the current professional structure of mathematics and its dynamic character were generated only recently. Whether the image of the mathematician as a professional existed earlier or not, and if so, of what sort it was, remains for the present an open question.

Modern sociology understands "professionalization" as a social process in science which begins somewhere in the 19th century[1]. Accordingly "professional mathematics" cannot be expected before the 19th century. However, one cannot deny the existence of some form of "professional activity" before the 19th century. Needless to say, the professional forms identifiable in the 19th century are in various ways connected with previous occupational forms.

How should we proceed in this situation? As a minimal condition a professional structure of whatever sort must establish the outlines of an educational framework, an activity exercised predominantly through the methods acquired within this framework, and the financial means allocated for the support of this activity.

Were such a "professional mathematician" to be found in the seventeenth century, then in terms of our present conception he should be sought among the outspoken exponents of the new mathematics. But although the seventeenth century has appeared as a "century of mathematics" to later observers, among the creators of the new mathematics scarcely a single one meets the criteria sketched above. As a result of their social origins a portion of these men were spared the necessity of having to exercise a profession in order to supply their means of subsistance; others were lawyers, theologians, diplomats, or politicians, and conducted their mathematical researches as a recreational activity. Nevertheless a substantial number devoted some part of their lives exclusively or principally to mathematical problems and earned their living mainly through this means. This group includes, for example, Desargues (1591-1661), Cavalieri (1598-1647), Wallis (1616-1703), Mercator (1619-1687), Huygens (1629-1695), Barrow (1630-1677), James Gregory (1638-1675), Newton (1642-1727), Jakob Bernoulli (1654-1705), and his brother Johann (1667-1748). Seven of these ten held chairs for mathematics at a university, of which five were newly founded. If we take into consideration that Huygens was paid from the

Royal purse for his activities as an astronomer, physicist, and mathematician at the Paris Academy, then it appears that at least eight of the ten meet part of the criteria set forth abov It is worth noting, however, that each of these cases of a finan cially supported mathematician, ι :h the exception of Cavalieri occurs in the second half of the seventeenth century.

If one considers that in the sixteenth and first half of the seventeenth century chairs for mathematics were already in existence, housed in the arts faculties, then the preceding may lead us to suspect that if extraordinary performance in mathematical research played any role whatsoever in the appointment to such chairs, then it certainly did not do so until the second half of the seventeenth century. This could indicate that the search for some sort of professional self-image among the mathematicians of the seventeenth century is completely misguided, since the possibility of being introduced to research within the framework of a thoroughly established program of education theoretically did not exist until the end of the seventeenth century and was only practically introduced after the French Revolution. Our search for a professional image which is stamped by our present conception of the dynamic development of education and opportunities for employment has produced a conjecture which leads to a series of other questions: If mathematical knowledge at the beginning of the seventeenth century started from a relatively static condition due to the absence of any concept of research, perhaps a professional self-image oriented along the lines of the contemporary artisan may have provided a model. Did the designation "mathematician" as such exist, and if so, is this to be conceived as the designation of a professional? If there existed some concept of mathematical research, who was responsible for it, and how were its relations to a more or less fixed body of mathematics regulated? Is some entrance to mathematical research imaginable and perhaps even demonstrable other than that based on training regulated by an institution such as a university? Did the understanding of the nature of mathematics itself change during the seventeenth century, perhaps creating a shift in the professional image of the mathematician or an aspect proper to a later image of the mathematician as professional?

In order to answer these questions it is useful to construct a sketch of the understanding of mathematics around 1600, which, considering the varieties of opinion represented in the available sources, can only be a gross generalization of the facts:

The main function of the universities in the later Middle Ages and early modern period was to train future generations for professional activity in the state, the church, or medicine. As a necessary preparation for this end it was required to complete a course of study in the arts faculty, which included instruction in elementary mathematics. There existed no reason, therefore, to demand special qualifications of the teachers for this curriculum, which was stable for many generations; this would have only endangered that very stability which was itself regarded as the guarantee for the standards of education. As a result of this stable situation, the universities were able to recruit the personnel for the arts faculty from their own ranks; that is, suitably inclined and talented graduates of this course of study were immediately installed in the arts faculty after passing their exams. In addition, outside the university there existed groups of reckoning masters partially organized in guilds, which transmitted the necessary knowledge of basic techniques needed for business calculations, which likewise remained stable for centuries. In the early modern period a third group emerged, the mathematical practitioners, who had competence in areas such as geodesy, fortification, astronomy, artillery, navigation, and the production of the instruments proper to these concerns[2]. Without going into the potential overlap of two or all three of these areas in a single person, it suffices to point out that the stable character of these forms, particularly the first two, is their dominant feature. So far we have discovered that a professional image of a mathematician existed within these three groups. But the mathematics involved was not a science, it was an art. This is to say that it was static knowledge, a fixed repertoire of skills to be used and applied in established situations. Later when the idea of research made mathematics a dynamic field, where new methods were consciously sought and encouraged, the old social structure of mathematics was first altered and eventually replaced entirely. In what follows I will be concerned with this process.

Changes were introduced into the structure of mathematics through the availability of almost the entirety of the recovered sources of Greek mathematics in the printed editions of the sixteenth century. On the one hand a rich field of activity was opened up outside those unchangeable areas partially locked into the social forms of the artisan tradition. On the other hand this work was interpreted as the reconstruction of knowledge which had already existed; that is, even the discovery of new solutions to new problems within the domain established by Greek mathematics were to be attributed to the Greek tradition, only a small fraction of which had been transmitted. Research, no matter by whom it was conducted, was thus considered at first only as re-discovery. Confidence in the possibility of ever going beyond the bounds set by Greek mathematics was first awakened only about 1600.

At the level of the universities these developments produced few if any changes. An expansion of the course offerings in mathematics is seen in the foundation of new chairs of mathematics; but the example of Petrus Ramus in Paris illustrates that the principal work of the occupant of a chair corresponded closely to the educational function of the university, that is, it was limited to the rearrangement of, for example, Euclid's Elements according to the pedagogical standards of the time. His aim was to select materials useful for craftsmen and to make them understandable for thirteen-year-old students.

In the case of the reckoning masters these developments led principally to the creation of a new market in the form of nobles and powerful financial magnates, who in addition to elementary methods of calculation desired entrance into the reopened world of Greek mathematics. Moreover a new form of competition was introduced to determine the relative ranking among the reckoning masters which served as an index for orientation within the new market. These competitions consisted in challenging a colleague to solve a series of mathematical problems within a limited period of time. At the end of this period the two competitors met, frequently before a large audience, where the challenger was required to produce the required solutions, and in the case that he could not, the victor was to reveal his own solutions. There emerged among the leading reckoning masters of the sixteenth century, therefore, a strong demand to fashion for oneself a method which was at once capable of solving the problems posed by a competitor while at the same time permitting the formulation of the most difficult problems imaginable for one's opponents. This demand for guarding the secrecy of such newly developed methods corresponded to the similar pattern of monopolizing recipes and methods of production in the guild tradition. This secrecy secured the possibility of a steady income, but it stood in direct conflict with the growing demand for information, which was satisfied without restriction at the elementary level offered at the universities as well as at the elevated level offered by the Greek classics in the original and in translation.

The victory of unlimited access to information had two sources. First, the knowledge of the newly acquired theory of equations kept secret for some time by the reckoning masters seemed modest in comparison to the influx of knowledge from Greek sources and the widespread efforts to add to them. Secondly, those concerned with the solution of new problems in the domain established by Greek mathematics were not motivated by securing a source of income. For them the problem domain opened by the Greeks served as a stimulus for a new, creative intellectual arena in which the chivalric virtues of dexterity, technical skill, power, and endurance were replaced by the virtues of knowing about problems and their methods of solution, synthetic ability, and individual genius. The prize remained the same: success in this arena of combat brought materially incalculable growth in reputation and honor. It was the time when one strove to be a new Archimedes,Apollonius, or Euclid, a time when Adriaan van Roomen, the Apollonius Belgus, damaged the honor of France, which was repaired by François Viète, the victorious Apollonius Gallus. On this level qualities such as practical applicability and general utility did not play a role, for the neoplatonic tradition denied them the status of characteristics proper to pure mathematics, and at least some of the new Apollonii concerned only with their personal honor or that of their country maintained a conscious distance from the material utility of their favorite pastime.

In this situation only a loose relationship could be established between the amateurs responsible for the development of the new mathematics and the representatives of a traditional mathematics circumscribed by its own problems and methods of solution, such as that practiced in the universities, by the reckoning masters, and the mathematical practitioners. All of this changed slowly to be sure with the discovery that the methods found through the reconstruction of Greek mathematics could be employed with great success in the classical applied domains of mathematics, astronomy and mechanics. The developments originating with this discovery formed the basis of a new conception of mathematics which was reinforced by the value it assumed within the ideology of the pedagogical reform movements of the Protestants and Jesuits.

These different trends united at an early stage in the life of Johannes Kepler. Moreover, Kepler was quite conscious of the status of his profession, as mathematician-astronomer[3].

Even within the still imprecise image projected by the representatives of the various groups earlier designated as mathematicians a pronounced tendency to limit the connotation of the word "mathematicus" is visible by 1600. Thus the algebraical problem of solving an angle-section equation of degree 45 proposed by Adriaan van Roomen in 1593 was directed explicitly to "mathematicis totius orbis." Viète, who was not included by van Roomen among the fifteen professional mathematicians capable of solving the problem, emphasized in his own solution that he did not consider himself a member of the group of professional mathematicians, but rather of the amateur mathematicians who concerned themselves with mathematical problems in their spare time out of pure pleasure[4]. Viète's modesty is only apparent: he intended to have nothing to do with the stagnant mathematics of the practi tioners; rather he aimed at shaping a new mathematics according to the standards of rigor and purity of Greek mathematics. That such an approach did not succeed initially, that others, such as Kepler for example, had no interest in occupying themselves with Viète's algebra, was due to the large spectrum of possibilities for going beyond the rediscovered works of the Greeks; but even more importantly it was due to the non-binding character of the new mathematics. Precisely because a large percentage of the mathematical achievements of the seventeenth century was accomplished by amateurs, the usual forms of reward and punishment associated with professionalism were not operating. For those amateurs who had either another profession or no profession at all, it was of little consequence whether others valued the results of their mathematical hobby[5]. The same held for those who were active as mathematicians at universities, as practitioners or as private instructors; for the static field in which they worked was nearly independent of the open mathematics leading to new frontiers. An excellent example of this point is provided by Roberval, who held the chair of mathematics in Paris. The various disputes with Fermat and Descartes, in which he often exhibited crass misunderstanding and committed numerous errors[6], did not at all damage his position in Paris. A counterweight to this situation dominated by lack of unanimity concerning the nature and significance of the objects under investigation was offered by the unifying power of the standards taken over from the Greeks. Thus one could agree upon the correctness of a result, simplicity of a solution, or a demonstration, and that one ought to explore the greatest possible range of validity for a method.

With these standards the possibility was offered for distinguishing the significant from the less significant; and in addition, the possibility of application - or even better utilization - of the new methods within other branches of mathematics was offered as an aid for orientation. A strong hedonistic element played a central role in this "hobby mathematics", for the great degree of freedom resulting from the absence of controls as well as the fullness of open problems promising success led to a tremendous increase in the number of amateur mathematicians and an expansion of available knowledge.

Accordingly young mathematicians quickly encountered difficulties in attempting to link their work to the results of the new mathematics because the division between the static areas represented by the professional mathematicians and the dynamic research front explored by the amateurs was growing ever wider. This division could at first only be bridged by self-education or later through private instruction by someone who had contributed to the new developments. In the main this was due to the fact that it seemed impossible to change or expand the curricula of the Latin schools and the universities.

Symtomatic of these problems was the _Idea matheseos_ published in 1650 by John Pell. This book, which was based on the pedagogical ideas of Comenius, starts from the assumption of the utility of the new mathematics and from the necessity for private study as the only possible means of acquiring entrance to it. In order to meet these needs Pell suggested a plan for organizing autodidactic study involving the foundation of a public mathematics library, the construction of appropriate introductory texts with methodical guidelines for individual study, and a critical bibliography of the relevant literature. The public library was to collect not only all books in mathematics but also every mathematical instrument. In addition the growth of mathematical knowledge was to be registered in an encyclopedia, conceived as a sort of data bank for the new mathematics. Similarly, compendia and appropriate guides for individual study were to be published[7]. However utopian certain aspects of Pell's plans may have seemed at the time (although they were in fact applauded by Descartes and Mersenne[8]), they still clearly entertained the possibility of collecting the entire domain of new mathematical knowledge and giving it a coherent and unified presentation.

Pell is representative of those calling for the construction of an educational and organizational framework designed to serve as an entry to mathematical research. But such appeals had only a limited success. Although the idea of a central mathematical library was never realized, a portion of the new mathematical literature was purchased by the university libraries, particularly in England, thereby opening the possibility for individual study outside the standard curriculum, which continued to outlaw the new mathematics. On the other hand, the demand for including the new mathematics within the university curriculum required not only a change in the understanding of the educational task of the universities but also a higher degree of maturity of the new mathematics itself. Decline in the hedonistic appeal of mathematics was a necessary condition for these developments as well as increasing unanimity within a recognized scientific community concerning the object, goals, and value of mathematical research. The achievement of this unanimity was still an object of struggle within the developing scientific societies in the second half of the seventeenth century, which from the very beginning attempted to separate experimental and theoretical natural philosophy and mathematics from rampant dilletantism. The correspondence of men like Barrow, Collins, James Gregory, Newton and Wallis from the sixties of the 17th century on shows clearly a growing awareness of the existence of an invisible college of mathematicians, being part of the often quoted "commonwealth of learning"[9]. This is connected with the development of a common terminology, criteria for distinguishing important from unimportant, good from bad. In addition we find increasing concern with making money as a mathematician[10], a "mathematicus". The connotation of the word mathematician or the Latin "mathematicus" seems to be unchanged up to the 19th century. Only in the second half of the 18th century in France did the proper designation of higher mathematics become "haute géométrie" and accordingly a research mathematician have to be entitled as "géomètre"[11].

The separation of a more elevated level from dilletantism required the creation of something new, namely, social forms for professional research in natural science and mathematics. In addition to ability and education, professional engagement in research obviously required the availability of time. If the circle of future scientists and mathematicians was not to be limited to those who could live from their own means, then possibilities would have to be found for compensating these new full-time researchers for their work just as in any other occupation. In the Academy of Paris, therefore, and later in the European academies that followed the same model, absolute governments allocated funds to support full members of the academies. These persons were in turn required to follow prescribed statutes in working together on commonly designed research projects, in acknowledging control over the quality of their results and in their activities as referees. In the Royal Society of England, where the members were supported by the good will of the king, but not financially, and therefore were required to pay membership fees, only the functions of quality control of scientific research and scientific communication could be fulfilled. Already by the end of the seventeenth century, however, membership in the Royal Society was sufficient qualification to insure success in finding a position at a university, in one of the many private institutions, or as a private tutor. In England and to a certain degree on the continent, the private tutors assumed the role of providing the preparatory instruction needed for participation in mathematical research, which now could be evaluated according to commonly accepted standards; the private tutors represented a special group that had evolved from the practitioners and reckoning masters. The ability to assume the educational function for the dynamic part of mathematics obviously presupposed participation in the actual process of research. Examples of such private tutors can be found from the end of the seventeenth century on among the group of Huguenots who fled to England, as for instance Abraham de Moivre[12]; but even James Stirling and Thomas Simpson temporarily belonged to this group.

These two forms of professional activity, continuing in the eighteenth century, brought to the rapidly expanding field of mathematical research a series of introductory texts in the most attractive areas of mathematics. They had a form similar to text books which utilized, for example, the formulation of unsolved problems as material leading to mathematical research.

In the course of the eighteenth century a degeneration of these two forms of professional activity is observable. This is particularly noticeable in the case of private tutors in England: for, following the patterns set earlier by the practitioners and reckoning masters, the attempt was made to separate off parts of the new mathematics and to specialize in this newly restricted area. One possible reason for this is that most of the students who sought a tutor were not interested in active research but rather wanted only to acquire an understanding of the major lines of development in an easily digestible form. The introductory texts were accordingly altered into textbooks, with the result that contact with research was lost, a situation symptomatic of the decline of mathematics in England after Newton. This development was carried further by the mathematicians in the academies of Europe; but here many additional factors led to the search for other ways to professionalize mathematical research. The few who were paid for their activities as mathematicians in academies were gradually limited by the increasing number of commissions for research into particular problems. At the same time multiple membership in several academies led to a complete internationalization of mathematics[13] and the concentration of control within a small group of the most capable mathematicians who were therefore able to determine the direction, content, and methods of research, as for example through the prize questions of the Paris Academy. The result of these developments was an ever growing alienation between mathematics in the universities and that in the academies, and by the end of the eighteenth century a dessication of analysis which had been exhausted through its applications to astronomy and mechanics. Accordingly, in the correspondence between d'Alembert and Lagrange the problem of how to develop research mathematicians is repeatedly discussed, and the future of mathematics is pessimistically compared with that of Arabian studies[14]. This situation was aggravated by the low prestige still accorded to a "savant" or a "géomètre" for a member of the aristocracy[15].

Euler also found this to be an extremely pressing problem after the attempt was made to gradually fill the Petersburg academy with native Russians, replacing the foreigners originally called to the Academy. In Berlin Euler sought to train young Russian mathematicians through private instruction, carefully introducing them to research along with his son Johann Albrecht[16]. The method consciously used by Euler of systematically increasing the demands on performance until his charges were capable of independent mathematical research had validity in the nineteenth and twentieth centuries as well. His successful approach marked the beginnings of a Russian school of mathematics. For the solution of the problem of how to generate creative mathematicians in large numbers, however, the achievements of Euler alone were not sufficient.

In order to bridge the gap between the research-oriented academies and the teaching-oriented universities, paths were cut in the eighteenth century. Thus the professors at the University of Göttingen, founded in 1737, were obliged to do research as well as teach. For the eighteenth century at least, the new emphasis on combining research and teaching at Göttingen brought no noteworthy advances in mathematics. The professors available to respond to such a call were not at once in a position to become researchers simply on demand. Moreover, in those cases in Göttingen where research was actually conducted, a corresponding connection to teaching could not be made so abruptly. The viewpoint of organizing the curriculum in order to fit research to teaching emerged in France during the last years of the eighteenth century and was first realized in Prussia during the first half of the nineteenth century. These developments concern already the nineteenth century for which I want to formulate some presuppositions in a sketchy form.

One precondition was the expansion of the educational function of the universities in various respects. In France at the turn of the century, the awareness that academic mathematics could be applied to technology, which by itself had gained importance in economics and military sciences, prepared the ground for raising considerably the position of mathematics within the general educational system. In Prussia this awareness was complemented by a new ideology stemming from a neohumanistic movement which secured a new form and status for pure mathematics[17]. This meant that the social strata from which were drawn those who were able and even forced to concern themselves with mathematics were extended far beyond the privileged classes. The fact, for example, that from at least the middle of the eighteenth century every French officer was required to demonstrate a command of practical mathematics, also meant that in the nineteenth century mathematics no longer needed to depend upon the framework of preparation for the higher faculties but could now be studied for its own sake. Connected with this was also the fact that the researchers earlier active primarily in the academies were brought into the universities as teachers responsible for carrying out research. In addition it meant that gradually a complete educational system preparatory to university study was created and fitted to the new circumstances. As a result of this development the practitioner and the private tutor disappeared in the nineteenth century and with them the model of a professionally self-contained mathematics. In its place came professional activity primarily as teachers in Gymnasia or universities engaged in a mathematics dynamized through the connection of research with teaching. In the course of the nineteenth century ever greater domains of activity were opened for mathematicians trained in the universities.

Notes

1. Nathan Reingold in "Definitions and Speculations: The Professionalization of Science in America in the Nineteenth Century", The Pursuit of Knowledge in the Early American Republic (eds. A. Oleson and S.C. Brown), Baltimore-London 1976, pp. 33-69, holds that this process started in the U.S. in the 1860's. This would have occured several decades earlier in Europe.

2. See Ivo Schneider," Die mathematischen Praktiker im See-, Vermessungs- und Wehrwesen vom 15. bis zum 19. Jahrhundert", Technikgeschichte, 37 (1970), pp. 210-242.

3. See Kepler's letter to an unknown lady written in 1612, where he discusses the prestige connected with the title "mathematicus" or "Sternseher", in: Johannes Kepler, Gesammelte Werke, vol. 17, München 1955, pp. 39-44, esp. 40f.

4. See F. Viète, Opera mathematica, Leiden 1647, p. 305.

5. See Michael S. Mahoney, The Mathematical Career of Pierre de Fermat 1601 - 1665, Princeton 1973, pp. 20f.

6. Compare Ivo Schneider, "Descartes' Diskussion der Fermatschen Extremwertmethode - ein Stück Ideengeschichte der Mathematik", Archive for History of Exact Sciences, 7 (1971), pp. 354-374.

7. See E.G.R. Taylor, *The Mathematical Practitioners of Tudor and Stuart England*, Cambridge 1954, pp. 81f.

8. See Taylor *ibid.* p. 216.

9. See *Correspondence of Scientific Men* (ed. S.J. Rigaud) in two vols., Oxford 1841, esp. the letter from Collins to Baker on August 19, 1676 or Cotes to Jones on February 15, 1711.

10. See especially Collins's letters to various persons in Rigaud *ibid.*

11. This becomes especially clear in the correspondence between Lagrange and d'Alembert. See *Oeuvres de Lagrange*, vol. 13, Paris 1882.

12. See Ivo Schneider," Der Mathematiker Abraham de Moivre (1667-1754)", *Archive for History of Exact Sciences, 5* (1968), pp. 177-317.

13. This is beautifully illustrated by the letter from Johann Bernoulli to Euler on April 23, 1743, in which he contends that he would leave his country for the sake of better conditions for mathematical research.
See *Correspondence Mathématique et Physique de quelques célèbres géomètres du XVIIIème siècle* (ed. P.-H. Fuss), in two vols., Petersburg 1843, esp. vol. II, p. 524.

14. See letter from Lagrange to d'Alembert on September 21, 1781; _Oeuvres de Lagrange_, vol. 13, p. 368.

15. See e.g. letter from d'Alembert to Lagrange on April 10, 1769, where d'Alembert reports that Condorcet's family has given up their resistance to his membership in the Academy even though many aristocrats believe "que le titre et le métier de savant dérogent à la noblesse". _Oeuvres de Lagrange_, vol. 13, p. 130.

16. See _Die Berliner und die Petersburger Akademie der Wissenschaften im Briefwechsel Leonhard Eulers_ (ed. A.P. Youschkevitch and E. Winter) in three parts, Berlin 1959, 1961, 1976

17. See Gert Schubring's article in this volume.

THE CONCEPTION OF PURE MATHEMATICS AS AN INSTRUMENT IN THE PROFESSIONALIZATION OF MATHEMATICS

Gert Schubring

J.D. Bernal characterized the growth of science during the first half of the 19th century by the following paradox: "At the time when science should have been most obviously connected with the development of the machine age, arose the idea of pure science". (Bernal 1973, 29) This paradox of the relationship between the dimension of development and application of science may serve to provide a better understanding of the problems of the professionalization of scientific activity in mathematics.

The transition from the 18th to the 19th century meant for mathematics the beginning of a fundamentally new phase in its development. The most prominent feature of this new development was that the type of mathematician usual at that time - the amateur, the practitioner, the universal scholar - was superseded by a full-time researching and teaching mathematician. In the course of this process a specific place for mathematical activities emerged - in contrast to the wide variety of locations which had existed previously - : the university. At the same time, mathematics developed an autonomous communication-network: more and more research activities became a matter of continuous development, following selfimposed aims, so that external impulses such as Academy prize questions, etc. were no longer needed.

Communication became increasingly down-to-earth. Under the constraints of the subject, former excesses such as pointing out one's own merits and trying to humiliate one's competitors were reduced, which meant that the regulation of the method through the content increasingly prevailed.

The notion of discipline and professionalization

It is clear that these qualitatively new developments suggest a description in terms of "discipline" and "profession". To be able to use these terms sensibly for the further explanation of this new course in the development of mathematics it is necessary to explain them.

While sociology has been analysing professions intensively for decades, there is a lack of comparable work concerning "disciplines" due to the fact - according to Stichweh - of a "lag in the sociology of science" (Stichweh 1980, 1). The traditional notion of both terms is that of a separation according to cognitive and social factors: whereas "discipline" means the unity of science as far as knowledge is concerned, "profession" embraces the social dimension of science. In connexion with the discarding of the one-sided view of science as a system of knowledge - the overcoming of the "statement-view" (cf. Stegmüller 1974, 172) - and the increasing interest in the social history of science, historians of science frequently fell back on the term "profession" in order to include the social component. Without "establishing sufficient contact to sociological theory" (Stichweh 1980, 1) simply by taking over the term, the concept of profession was often reduced to "full-time and remunerated employment" (Crosland 1976, 139).

Obviously the adoption of these terms as they were developed by sociologists for the analysis of traditional professions such as doctor, clergyman, lawyer, etc. failed to take sufficient account of processes connected with the institutionalization of science. Stichweh therefore criticized the additive use of "profession" to include social factors, since it implied that "the processes of scientific communication ... are basically non-social processes". He therefore proposed to take the term "discipline" as basic for the study of individual branches of science, and to regard it as an "integrating term which embraces both cognitive and role-oriented relationships"

(op. cit. 2). Such a unity of social and content factors can be found in the characteristics of disciplines as given by Stichweh, who defines them as "forms of social institutionalization ... of processes of cognitive differentiation in science":
Sufficiently homogenous communications between researchers, i.e. a "scientific community"; a stock of theoretical knowledge represented in textbooks, i.e. characterized by codification, acceptance by consent, and basic teachability; a plurality of problematical questions at any time; a "set" of research methods, and paradigmatic problem solutions; a discipline-specific career pattern and institutionalized socialization processes which serve to select and educate candidates according to the prevailing paradigms (R. Stichweh 1979, 83).

Indeed, it was just the relationship between content and social factors which had characterized scientific activities since the 19th century. While it seemed to be justified in the 18th century, to regard disciplines as systems of knowledge and to examine the role-relationships of scientists and society separately, in the 19th century disciplines gained a new character with the emergence of the modern branches of science and their institutionalization, which was manifested in the unity of social and cognitive factors. In the present article professionalization of science is thus defined as the process of the emergence and ultimate domination of the unity of cognitive and social factors within a scientific discipline.

One could of course ask whether the term "profession" ought not to be reconstructed analogously to the term "discipline". As a matter of fact, it is not possible to understand the "Verberuflichung" of scientific activities unless the traditional sociological categories are linked with subject characteristics of the kind mentioned above, as will be explained below. Stichweh, however, regards the term "profession" as not generally applicable to sciences, as there are different relationships between society and profession on the one hand and profession and disciplines on the other: whereas the typical professional activity of a member of a traditional profession is complementary to a non-professional client, and thereby characterizes "profession" by the application of knowledge, scientists are interested in creating new knowledge; scientific disciplines tend to be inward-looking rather than interested in active relationships with the world around them. Professional associations are therefore more concerned with protecting themselves from interference and external control than are disciplinary communities and associations, which are more oriented towards internal communication (cf. Stichweh 1980, 3-6).

It seems that this differentiation - and above all the relationship to application - only concerned a particular phase of scientific activity. But it is certainly true that the low militancy of disciplinary associations reveals a remarkable fact: the institutionalized disciplines had been safeguarded up to that point by the social system in such a way that they did not need any extensive external representation. The academic disciplines had thus achieved a degree of social recognition which professions and their associations were still striving for. The analysis of the development of this specific relationship between science and society constitutes an even greater task.

Since the institutionalization of a new discipline has been regarded as particularly important for the study of "the relationship between intellectual and social processes" in scientific development (Lemaine et.al. 1976, 17), I will consider the institutionalization of pure mathematics as a particular feature of that process described above as professionalization. Acceptance of its legitimacy is crucial for the establishment of a discipline (Lemaine et.al. 1976: 17; Thackray/Merton 1972: 473). I will first consider some dimensions of the process associated with establishing legitimacy. To achieve acceptance it is essential for a discipline to establish the relationship between social recognition and the development of a methodology.

Thackray and Merton have stressed the importance of "metaphysical assumptions, and particular Weltanschauungen" for the emergence of a discipline (loc.cit., 474). A methodology - which is also very much determined by the subject - not only has the task of securing the coherence of a discipline and the unity of methods and problems, but also faces the task of justifying itself - towards other disciplines as well as society as a whole. It is therefore necessary to have an agreement between the disciplinary methodology and the prevailing system of values - the Weltanschauung - at least for the period of institutionalization.

So far, the sociology of science has not sufficiently taken into account the fact that there is one factor which is vital for social recognition: the state. It is indeed true that the establishment as well as the continuous development of a system of science requires decisions and means which are not available on the basis of sectional social interests. Such institutions of a social sub-system require a continuity which only the state can provide, as it acts on behalf of society as a whole. At the same time, it is evident that the state - in order to take over these functions - has to override interests which are too narrowly bound to particular sections of society. It is not surprising therefore that the professionalization of science did not begin until the feudal state had been more or less superseded and the middle classes had risen to a strong position in the state (cf. Schubring 1980 a). The necessity of social acceptance for a discipline by those who act on behalf of a nation's society is reinforced by the fact that the new social sub-system cannot exist in its own, but needs a sub-structure such as would be most effectively provided by a general education within the school system. If it is true that the attention of disciplines and university systems and their associations are mainly centred on their internal affairs and therefore less oriented towards resistance against intrusion and towards gaining control, it suggests a high degree of legitimacy. Proof for this close relationship can be found in a negative example: after the end of the 19th century French governments more and more removed the "Facultés des Sciences" from state control which led to a significant decline in professionalization (cf. Shinn 1979, PP. 314 - 326).

It is important, therefore, for the professionalization of mathematicians' activities, to analyse the relationship between the meta-conception of pure mathematics and the general methodology of sciences then prevailing in Prussia. R.S. Turner has shown that there was such a methodology in Prussia in the first half of the 19th century, which he analyzed as "Wissenschaftsideologie". He pointed out that this "Wissenschaftsideologie", which is also known as "neo-humanism", was the common scientific ground - "Weltanschauung" - of the leading reformers: those in influential position in the Government, the scientific and educational fields (c.f. Turner 1973). This served the government - especially the Kultusministerium - as a basis on which to mould universities and schools to the new structure. At the same time the "Wissenschaftsideologie" - connected with a reform of learning - was part of a more dramatic reform of society and the economy: the use of education for the encouragement of economic activity-"industriousness" ("Gewerbefleiß"). This policy was intended to facilitate the transition from the primarily state-organized economic activity of the prevailing mercantilism, to an increasing mobilization of private initiative, a prerequisite for middle-class modes of production.

Furthermore, Turner pointed out that the "Wissenschaftsideologie" led to the emergence of the "research imperative" for scientific activities. It was only the development of the "research imperative" at the reformed Prussian universities after 1810 which established the "dual role" of the teacher-researcher as the characteristic feature of the institutionalization of scientific activity (cf. Turner 1973). As far as the professional role of scientists was concerned this dual activity became the significant content feature of the newly emergent social role. For example, levels of qualification were laid down, which led to certain career patterns as a specific feature of the professionalization in the field of science.

On the other hand, the development of the "research imperative" and of professionalization at Prussian universities, as shown by Turner, concerned at first only philology. Textual criticism was the newly established research methodology: it was governed by the paradigm of the unity of justifying or systematizing and disseminating knowledge, and not on a unity of developing new knowledge and disseminating it as a prerequisite for its pervasive social application: "a type of academic originality and research which served the ends of synthesis, not analysis" (Turner 1972, p. 155).

It is well known that the methods of philological science served as a model for mathematics and science. Thus C.G.J. Jacobi took the classical philology seminar of his teacher Böckh as his model for educating scientists in mathematics. Mitscherlich, who also started his scientific career as a philologist, is reported to have developed the chemical concept of isomorphism along the lines of comparative language research (Lenz 1910, p. 226).

Mathematical Methodology

Until recently, however, it was not known how the transition to subject-specific methodology, which is necessary for professionalization took place in mathematics. I shall go on to discuss this transition with regard to the autonomy of the discipline. The emergence of the autonomy requires the development of subject internally defined objectives and values, and, by the same token, methods.

A recent case study concerning the plans to establish a polytechnical institute in Berlin between 1817 and 1850 has shown that the main function of these plans was to effect the professionalization and institutionalization of mathematics as an independent, modern science, particularly through the creation of full-time positions for mathematicians which would have enabled them to do research and teaching in mathematics on the basis of living salary (Schubring 1979).

As the university during the twenties of the previous century had not yet become the institution suited for this purpose, the polytechnical institute was to be established separately. The plans acted, however, as a means to transform the Prussian university in such a sense as to make the "research imperative" a constitutive factor for mathematics and the sciences there as well.

The autonomy of mathematics as a discipline required not only the development of subject-specific methods, but also in a certain sense the pursuit of pure mathematics as an end in itself, as we know C.G.J. Jacobi to have done. Jacobi, whose most important motive was the professionalization of mathematics as a university discipline, was concerned with securing its autonomy. This aspect sheds a new light on his rejection of the externally defined value of usefulness, and on his corresponding emphasis on the internal values of the discipline.

In the empiricist conceptions of science held by the English and French materialists of the 18th century, the relationship between development and the application of knowledge had not been recognized as a problem. True, they distinguished between pure and applied mathematics (using different designations), but they did so more in the sense of systematically distinguishing knowledge, and this was not meant to establish separate disciplines in the sense of autonomous communicative metworks (cf. Stichweh 1977).

The pessimistic suspicion - especially of French mathematicians in the second half of the 18th century - which feared that mathematics might come to a standstill, as e.g. Arabic philology had done, is well known. For the further development of mathematics it was necessary to have a methodology which accounted for the possibility of theoretical, new knowledge - in contrast to the then prevailing belief of the analytical method which said that propositions of pure mathematics were always more or less trivial transformations of the "éléments". Toth has shown that this situation was radically changed by the philosophy of Kant: Kant proved that there can be essentially new knowledge in pure mathematics - this is the real essence of Kant's claim for synthetical a priori propositions (Toth 1972, p. 8).

Only on the basis of Kant's epistemology, which was dualistic at the core (Buhr/Irrlitz, p. 40), did it become possible to reflect on the relationship between pure and applied mathematics. Kant had pronounced himself strictly against "mixing up" different epistemological principles ("reine Anschauung" with experience) and for a separation of the pure and the applied sections of the sciences (cf. Kant 1977). On the other hand this new start was initially elaborated more in abstract philosophical terms rather than subject oriented. This lack of subject orientation was criticized by Klügel in his well-known "Mathematisches Wörterbuch": "Kant behauptet, daß die Philosophie eben sowohl von Größen handele als die Mathematik." But the philosophical "Größen sind doch von einer anderen Art als die mathematischen". Klügel, however, also pointed out: "Das Wesen einer Wissenschaft beruht auf ihrem Gegenstand, und ihre Methode wird durch diesen bestimmt" (Klügel, vol. 3, 1808, pp. 620 sq.).

It is a fact in the history of mathematics which is almost forgotten today that - starting from the foundation established by Kant - the subject-specific methodology of mathematics was founded as "pure mathematics" by the philosopher and natural scientist (and quasicounterpart to Hegel) J.F. Fries (König/ Geldsetzer, pp. 44 sq.).

While separating the more fundamental philosophical prerequisites, the philosophy of mathematics asserted the reality of mathematical concepts and the certitude of mathematical results. Crelle drew on this new autonomy of the methods of mathematics, on its self-confident independence of philosophy, pointedly formulated by himself, to justify the Berlin plans to establish a polytechnical institute.

This philosophy of mathematics is in stark contrast to the philosophical programme of Hegel und Schelling, who wanted to subordinate the methods of mathematics to the methods of philosophy. Hegel for example declared that mathematics was "dead" and therefore incapable of any internal motion (Hegel, p. 157).

It is thus not accidental that no supporters of Hegel or Schelling, but rather adherents of Fries are to be found among the "modern mathematicians", particularly among the promoters of the Berlin plans. Gauss was very much in favour of Fries' philosophy of mathematics and thought of him as the only philosopher he could trust (Cf. Nelson, pp. 437 sq.). Fries' methodology was based on a sign conception of mathematics. Development and application are separated parallel to the relationship between sign and meaning. The fundamental discipline of pure mathematics is "syntactics" or "semiotics", as the general theory of mathematical operations. Fries considers the composition of equal elements the basic operation of mathematics This is the reason for the attractivity which the "Kombinatorische Schule" had for the elaboration of the programme of pure mathematics. At the same time, the radical separation of sign and meaning by Fries led to the introduction of variables. Fries did not confine himself to elaborating a philosophy of pure mathematics, but also developed at the same time a philosophy of applied mathematics. In order to connect these two, he called for a "<u>theoretische Lehre der Vermittlung</u>" (theory of mediation), which was to reflect the condition and aims of applying the 'core' of the theoretical structure with its initially empty abstractions, and signs devoid of meaning (Cf. Fries 1822).

The significance of such a meta-conception can only be explained further by a comparison with France. So far it has not been established whether mathematics too suffered that "decline" after the French Revolution or after 1830/40 which in recent years has been discussed under the name of the "decline of French science" (Fox 1973, Shinn 1979) as for example F. Klein maintained. (F. Klein 1926; see in contrast Shinn 1979). However, there is no doubt that in France, in contrast to Prussia, no institutionalization of mathematics took place, and no conception of an independent discipline "pure mathematics" developed.

In fact, in France no methodology had been developed which would have overcome the utility orientated empiristic conceptions of knowledge held by the French Enlightenment. Wronski, who was _de facto_ the only one in France to make an effort to develop an independent methodological discussion of the principles of mathematics, did this from a position of a complete outsider. But he, too, failed to understand the difference between development and application, and he was therefore unable to justify "pure mathematics" when he said for example, that _mathématiques pures_ analyse the same subjects " in abstracto", which _mathématiques appliquées_ enquire into "in concreto" (Wronski 1811, 3 bis). The small influence which Wronski's reflections had on the basics of mathematical thought is apparent, although Gergonne introduced the new section "_philosophie mathématique_" in his annals (beginning with Vol. 4, 1813) after the publication of Wronski's "Philosophie des Mathématiques", with the purpose of discussing the questions it raised. However, after some initial interest, only a few articles were published, mostly by Gergonne himself, so that this topic actually no longer appeared from 1822.

This lack of development of an explicit subject-specific methodology has to be seen in relation to the suppression of the scientific _Weltanschauung_ of the Enlightenment, which had been pursued since Napoleon, resulting in the absence of a general, socially accepted scientific methodology, in terms of which the various disciplines could have found their justification.

There was no scientific value system comparable to the "research imperative" of post-revolutionary France. This fact resulted in the remuneration of scientific activity being based almost exclusively on teaching. This, in turn, gave rise to the "cumulards": in order to survive many scientists were compelled to accept several chairs simultaneously (Cf. Fox 1973). This of course made the development of a scientific career for younger scientists increasingly difficult. The lack of scientists' orientation towards research as a value system, their continued adherence to the utilitarian positions of the Enlightenment or their transition to positivism, may be understood as a result of successive French governments' anti-science policy, which was quite consciously pursued from Napoleon's time onwards up to the second half of the 19th century. Sève has shown that Napoleon's objective was to eliminate those philosophers described as "idéologues", who, in the tradition of Condillac and Condorcet, continued to make a stand for reforming society by means of science. They were sucessfully supplanted by adherents of the philosophy of 'spiritualism', which had, in fact, been newly created with the support of the state, and which adapted the Scottish school of philosophy - (also called the philosophy of common sense) - for French purposes. 'Spiritualism', which allowed the clericals to reconquer the educational system, thereupon rose practically to the position of a state philosophy (Cf. Sève, pp. 18 sq.). Simultaneously, this served to effect a re-orientation in the leading social values from scientific to literary-political ones (Cf. Fox 1973).

The impact to mathematics

The meta-conception of pure mathematics supported the professionalization of mathematics activity in Prussia in several respects.

First, it made for the freedom from externally defined objectives of this discipline, and for a re-orientation towards discipline internal values: this was the precondition for a gradual shifting of the authority to appoint new scientists from a ministry intend on safeguarding the scientific level of the discipline to the representatives of the discipline themselves.

It laid the foundations for the acceptance of a dual teacher-researcher activity as a full-time occupation, the remuneration of which had to be sufficient to live off. The transition from laying the main emphasis on teaching, which compelled the teachers to seek additional part-time teaching posts, to a dual activity, in which teaching comprised only the lesser part of the remunerated activity, marks the decisive step towards professionalization.

The specialization connected with the rise in the conception of pure mathematics simultaneously made it possible to establish an education for scientists and a scientific career (with the stages of Doktor, Privatdozent, Extraordinarius, Ordinarius), which, in turn, enabled the discipline to channel and to reduce the number of self-taught scientists besieging the ministry and the faculties. It is significant that the education of scientists as outlined in the Berlin plans was to be carried out in the form of "Seminar", together with the education of schoolteachers: as long as the discipline itself had not yet developed a specific model for educating scientists by means of co-operative research activity, scientific training was modelled on the training of lecturers (Dozenten), which thus could be provided along lines similar to that of the other teachers. The two models for the "seminary" which was to be built up according to the plans for the institute, are typical of this unity between the two aspects of teacher and scientist: the philological seminaries of the Prussian universities on the one hand, and the training provided at the Paris École Polytechnique on the other. By the latter, however, was not meant the official training - the education of engineers - but the second, "hidden" education provided there: the training of gifted students as répétiteurs, and finally, as scientists (Cf. Schubring 1979).

It is crucial for an evaluation of the meta-conception of pure mathematics to realize that it did not stand in conflict with the social demands for knowledge. Rather, the very emphasis on the sign function of mathematical concepts is an expression of its orientation towards the new social requirements regarding the dissemination of knowledge. The close relations of the emerging discipline mathematics with society was mediated throu its being embedded in the prevailing _Wissenschaftsideologie_ of Neo-Humanism and through its "base" (substructure) of _Gymnasium_-teachers thereby created. The professional orientation of the _Gymnasium_-teachers as a whole, as well as the specific activity with regard to the contents of mathematics teaching finally had a positive effect on the development of the discipline.

It is known that after Humboldt's reforms _Gymnasium_-teachers saw themselves as scholars ("Gelehrte") and worked in a scholarly way. It is rather less commonly known, however, that this scientific orientation was a deliberate strategy on the part of the Prussian _Kultusministerium_ with the intention of cultivating a high social standing for teachers and of actively encouraging learning and the acquisition of knowledge among the middle classes. Two examples will be quoted of this ministerial strategy to rouse the teachers to scholarly activity. One example concerns the so called '_Conduiten-Listen_' in which the supervisory school authority had to report annually to the ministry on the behaviour of each teacher under the headings e.g. "administration", "moral conduct". From about 1819 two new columns were to be answered by order of the ministry: "Ob er fleißig fortstudiert" and "ob er ein Schriftsteller ist und was seine neuesten Werke sind" (compare for example ZStA I, folios 19 sq.). In 1824 this stress on individua scientific activity was extended still further by the ministry through a decree which obliged each _Gymnasium_ to publish a scientific treatise ("wissenschaftliche Abhandlung") within the so-called school programmes "on a scientific subject not foreign to the teacher's profession and suited to call the attention of at least the educated classes to matters of public education in general, or to a topic of interest to _Gymnasiums_ in particular." Year by year the headmasters and Senior teachers took it in turns to write the thesis, which had to be decided internally. The objective was to encourage them "to continue their studies without interruption" (Rönne, p. 155). By this means, the ministry had created an effective instrument of professional communication. However, this was only a first step towards disciplinary communication, because the mathematics teachers used the dissertation not foremost for active communication but for reasons of purely personal prestige. Particularly in the early stages the authors were very keen in pointing out the advantages of their own systems while concealing the tracks of communication, such as e.g. their relation with predecessors in the field.

The scholarly activity of the mathematics-teachers was the basis for the survival and further development of the "combinatorial school". The most interesting fact about this school is, however, that it only came to full fruition after the death of its founder Hindenburg in 1808. This can only be explained by the fact that as an "educational mathematics" it seemed to be an adequate conception for a training that aimed at the development of mental abilities through the acquisition of the _fundamental_ concepts of the sciences. The mathematics of the combinatorial school obviously had an analogous function in the _Gymnasium_ as set theory has lately had in our primary schools.

At the same time this "educational mathematics" was of great importance for the university discipline of pure mathematics, since the combinatorial school preached the importance of the formation of concepts quite regardless of their potential applicability and usefulness (Cf. quotations from J. Grassmann in: Schubring 1980 b).
The possibility cannot be discounted for example, that the _geometrische Combinationslehre_ initiated the breaking of the ties of geometry with three-dimensional space. Supported surmise may be found in a remark by Scheibert's, who wrote in one school-programme (which followed J. Grassmann's conception of mathematics, when he as well as J. and H. Grassmann were teaching mathematics at the same _Gymnasium_ in 1834): "Die Geometrie kann nicht einmal über Produkte von drei Factoren hinaus, indem

sie mit ihren Constructen an die drei Dimensionen des Raumes gefesselt ist." (C.G. Scheibert 1834, p. 13).

This orientation of "educational mathematics" towards the fundamentals, towards the elements of concepts in particular was itself supposed to ensure the full applicability of knowledge.

Pure mathematics, according to Crelle, was also to have the role of meta-knowledge, that is knowledge about knowledge and about its application. The basic social prerequisite and at the same time the prerequisite of educational theory for this conception of a mediation between development and application was a general demand for the dissemination of knowledge in society as well as the insight that knowledge could, in principle, be learned. Crelle, who argued for pure mathematics even more narrowly and more pointedly than Jacobi, in his plan for the institute of 1828 has formulated this social prerequisite as follows: "Everybody, without exception, needs pure mathematics". ("Der reinen Mathematik bedarf jedermann ohne Ausnahme", ZStA II, fo. 34).

This conception differs fundamentally from educational theories based simply on aptitude.

It should not be overlooked, however, that the professionalization of mathematicians which was generally sanctioned by the 1866 examination regulations requiring from all "clients" (students) of the mathematical discipline the ability to make scientific studies of their own, stood in a certain conflict with the future profession of most of the students, who were studying to become teachers.

It must be added that no disciplinary association of mathematicians developed during the first half of the 19th century, as is otherwise typical for the professions. Jacobi, however, as the "head" of the Königsberg School exercised a similar function by speaking for mathematics in Prussia and addressing demands for material equipment to the Kultusministerium. Its first seeds are seen in a mathematicians' meeting which "was held in the early thirties in Berlin" and in which "took part besides others C.G.J. Jacobi, Minding and the two brothers Ohm", as is reported in Cantor's call for the foundation of the Deutsche Mathematikervereinigung (Chronik 1890, p.3. I could find no documentary evidence for this report up to now.).

The relation of M. Ohm to the discipline

It is interesting to try test the applicability of the characteristics of the professionalization described above on a mathematician who stood in extreme contrast to the representatives of the new discipline mathematics at Prussian Universities and who characterized in many respects the transitional stage in the history of the discipline: Martin Ohm. Ohm is known because of his contributions to the fundamentals of algebra (Cf. Novy), but he produced no essentially new results in research. Apparantly due to the transitional character of his position he has always roused great interest within the history of mathematics (Cf. as newest publications: Dauben 1980, Mehrtens 1980). As it is clear that the state and especially the Kultusministerium played an important and active part in the process of professionalisation, it is interesting to trace Ohm's relation with the Kultusministerium. Actually an inspection of Ohm's personal files reveals remarkable indications which give a completely different picture to the so often quoted supposition of Lorey, that Ohm had influential supporters inside the ministry (Lorey 1916, p. 35).

Süvern's marginal note - "good intentions" - in Ohm's memorandum of 1818 for the improvement of mathematical education which only supports this guess, merely refers to a partial aspect of Ohm's plan, namely "eine mathematische Pflanzschule mit der Universität in Verbindung (zu) bringen" (ZStA III, fo. 6). Ohm's essential demand to exempt prospective mathematics teachers from the examination for higher teachers and only to make them take "an examination in mathematics and sciences" was severely rejected in a marginal note: the consequence of "Lehrer bloß für diese Wissenschaften zu ziehen", would also require "besondere Lehrer für Philologie u.s.f." to be educated and would therefore violate the necessary "pädagogische Einheit". Süvern remarks: "Ohm selbst ist ... einseitig gebildeter Mathematiker" (l.c., fo. 5).

Ohm's pedagogical efforts were not regarded as in accordance with the neo-humanistic conception of education. Consequently his career cannot be considered as in any way comparable with those of Jacobi, Dirichlet and others. Their appointments to professorships and further advancement were based on their extraordinary research achievements; but it seems that such a criterion was never applied to Ohm. Characteristic for this is the peculiar procedure that in 1820 the ministry subjected Ohm to in an examination, if he were actually to be taken into consideration for a professorship at a university, despite the fact that Ohm had worked as a University lecturer in Erlangen for several years. Ohm, who at that time was a _Gymnasium_ teacher in Thorn, had applied to the _Kultusminister_ for a "University post" and had handed in "Neun Thesen" on Euclid with the request to the ministry to organize a colloquium on these theses with the Academy of Sciences. The "Special Commission", chaired by E.G. Fischer, that was finally set up by the ministry, suggested delegating the further examination of Ohm to the _Wissenschaftliche Prüfungskommission_, - the appropriate board for teachers! - and recommended a mathematics examination before this board as well as a public lecture. The fact is that Süvern has, indeed, acted in favour of this extraordinary examination. Ohm had supposed that the demonstration lecture would take place in a university, but Johannes Schulze gave orders that it should be held in a _Gymnasium_. Although the _Wissenschaftliche Prüfungskommission_ declared after the written examination and the demonstration lesson that they were convinced that Ohm "sich zu einer außerordentlichen Professur der Mathematik bey einer Universität schon jetzt wohl eignen werde" (l.c., fo. 51), the ministry did not offer Ohm a chair. Instead, Schulze asked him to "habilitieren als ... akademischer Dozent bei der philosophischen Fakultät" in 1821 (l.c., fo. 77). The orientation to teaching without expecting any research achievements, which was connected with this position of an "academic lecturer", was clearly pointed out by the ministry after his appointment. Therefore Ohm was reminded in 1822 that he will have to prove himself to be a good academic lecturer. His qualification for a university career was seriously questioned and warned "Ihre vorgebliche Abneigung gegen eine Anstellung bei einem Gymnasio (zu) bekämpfen", and he should "nach einer reiflichen Prüfung Ihrer selbst erkennen...., wie sie vermöge ihres inneren Berufs und nach den bisherigen Erfahrungen mehr zu einem Gymnasiallehrer als zum Docenten bei einer Universität geeignet sind" (l.c., fo. 102). However, that did not prevent Ohm from constantly applying for a chair.

Many books which Ohm submitted were - in contrast with the practice with university professors - given to Gymnasium teachers for an expert opinion. There were generally no positive expert opinions on Ohm's work which could have supported his subsequent advancement. For his appointment as "außerordentlicher Professor" in 1824 he himself had handed in two references: the recommendation of a "Konsistorialrath Matthias in Magdeburg" and a work of E. Collins of Petersburg Academy, which referred positively to Ohms' publications (l.c., fo. 140 f). His appointment to a chair in 1839 can be related to external pressure: the Board of Studies of the United Artillery and Engineering School made an appeal to the Kultusminister to appoint Ohm Ordinarius (e.c., fo. 146). Ohm had evidently threatened to leave Prussia otherwise. It thus seems that the military directly put pressure on the king.

Ohm's scientific-pedagogical efforts also failed to meet the new disciplinary values of the exclusively technical reference but were directly focused on himself, as was typical for the university style of the eighteenth century. This is evident, e.g. from Ohm's 1828 request to the ministry to establish a "mathematisch-physikalisches Lehrseminar, nach meinen Prinzipien und unter meiner Leitung" (ZStA II, fo. 19). The ministry flatly rejected such a personalization of the organization of the seminar, which was to play such a key role in the ministry's conception of institutionalization of higher learning: it would such an institute "unter Ihrer Leitung nicht für nöthig und nicht einmal für nützlich erachten, vorzüglich insofern solches nach Ihren eigenen, bekannt gewordenen, und noch weiteren Diskussionen unterliegenden Systemen geschehen soll" (l.c., fo. 20).

Ohm's views of science were obviously rooted in the strongly personalized way of communication, and not integrated into the content-oriented disciplinary communication system. The above mentioned example of his challenging the academy to a disputation with him proves this as well as the numerous examples of his philippic against other mathematicians (cf. e.g. Biermann 1973). In order to obtain personal advantages, he did not hesitate to denounce other mathematicians. For example, in a memorandum of 1832 to the Bavarian ministry of the interior, he defamed Crelle as a quasi anti-German representative of a "französische Parthey" (party) in German mathematics, and declared himself the founder of a "German" "school of mathematics" (Bayerische Akademie, fo. 193). By means of this memorandum, Ohm attempted to win support in Bavaria for his plan of a seminar, in which teachers of mathematics and physics were to be trained "in diesen (his) Ansichten" (l.c., fo. 194). Crelle's criticism seems to be right, in that Ohm's plans of institutionalization would have effected, in a certain sense, privatisation instead of generalization of communication, in the discipline, and would hence have led to its conversion to a "guild": "es kann gleichsam eine zunftartige Schule entstehen, die wohl von allem, was die Fortschritte einer Wissenschaft hemmen mag, zu dem Gefährlichsten und Nachtheiligstem gehört (ZStA IV, fo. 14)."

He did not communicate with his colleagues via journals in the usual way: only late in his career some articles by him appeared in Crelle's Journal. His form of publication being books, which were still entirely under the influence of the methodological conception of a closed system of sciences prevalent in the eighteenth century, and which were intended to lay down the definitive foundations and the perfect construction of mathematics. He aimed at systematizing existing knowledge, not at gaining new knowledge.

Therefore, despite Ohm's undeniable merits in the systematization of elementary mathematics and in the dissemination of the idea of an operational approach to its fundamental concepts, Martin Ohm was in several respects at odds with the new processes of professionalization of mathematics. His method which rather adhered to the model of philological science and his way of doing things which was rather aligned to the educational or Gymnasium teacher type of mathematical scholar provide characteristical hints as to the transitional stages. Ohm's case elucidates the contradictions which had to be overcome in this process in order to make a unity of cognitive and social factors prevail.

REFERENCES

Primary Sources

Files of the former Prussian "Ministerium der Geistlichen-, Unterrichts- und Medizinalangelegenheiten" in the: Zentrales Staatsarchiv der DDR, Dienststelle Merseburg.

1. Rep. 76 IV. Sekt. 17 p, no. 1, vol. 1. (ZStA I)
2. Rep. 76 Vc. Sekt. 2, Tit. 23, Lit. A, Nr. 17. (ZStA II)
3. Rep. 76 Vf. Lit. O, Nr. 2, vol. 1 (ZStA III) and vol. 2 (ZStA IV).

Archive of the Bayerische Akademie der Wissenschaften, München. Akt VII, 51 (abbr. Bayerische Akademie)

Secondary Sources

Bernal, John D.: The social function of science. Cambridge/Mass. 1963

Biermann, Kurt-Reinhart: Die Mathematik und ihre Dozenten an der Berliner Universität 1810-1920. Berlin 1973

Buhr, M./ Irrlitz, G.: Der Anspruch der Vernunft. Köln 1976

Chronik der Deutschen Mathematiker-Vereinigung. In: Jahresberichte der Deutschen Mathematiker-Vereinigung. 1890, 3-14

Crosland, M.: The development of a professional career in France, in: Crosland (ed.): The emergence of Science in Western Europe, New York, 1976, pp. 139-159

Dauben, Joseph W.: Mathematics in Germany and France in the early 19th Century: Transmission and Transformation. In: Jahnke/Otte (eds.): Epistemological and Social Problems of the Sciences in the early 19th Century. Dordrecht 1980, pp. 371-400

Fox, Robert : Scientific Enterprise and the Patronage of Research in France 1800-1870. In: Minerva, 11 (1973), pp. 442-473

Fries, J.F.: Die mathematische Naturphilosophie nach philosophischer Methode bearbeitet. Heidelberg 1822

Hahn, R.: Scientific Careers in France in Eighteenth Century France, in: Crosland, op. cit.

Hegel, Georg W.F.: Sämtliche Werke. Jubiläumsausgabe in zwanzig Bänden. Vol. 6. Stuttgart 1956

Jahnke, H.N.; Otte, M.; Schubring, G.: Mathematikunterricht und Philosophie, in: Zentralblatt für Didaktik der Mathematik, 2/1977

Jahnke, H.N.: Zum Verhältnis von Wissensentwicklung und Begründung in der Mathematik - Beweisen als didaktisches Problem. Bielefeld, 1978. Series 'Materialien und Studien des IDM', no. 10

Kant, I.: Metaphysische Anfangsgründe der Naturwissenschaft. Werkausgabe, Band IX. Frankfurt 1977

Klein, Felix: Vorlesungen über die Entwicklung der Mathematik im 19. Jahrhundert. Vol. 1. Berlin 1926

Klügel, Georg S.: Mathematisches Wörterbuch. Volume 3. Leipzig 1808

König, G.; Geldsetzer, L.: Vorbemerkung der Herausgeber zum 13. Band von: J.F. Fries, Sämtliche Schriften. Aalen 1979, pp.17-94

Lemaine, G./Macleod, R./Mulkay, M./ Weingart, P.(eds.): Perspectives on the Emergence of Scientific Disciplines. The Hague/ Paris 1976

Lenz, Max: Geschichte der Königlichen Friedrich-Wilhelms-Universität zu Berlin. Zweiter Band, Erste Hälfte. Berlin 1910

Lorey, W.: Das Studium der Mathematik an den deutschen Universitäten seit Anfang des 19. Jahrhunderts. Leipzig/Berlin 1916

Lundgreen, Peter: Techniker in Preußen während der frühen Industrialisierung, Berlin, 1975

Mehrtens, Herbert: Mathematicians in Germany circa 1800. In: Jahnke/Otte (eds.): Epistemological and Social Problems of the Sciences in the early 19th Century. Dordrecht 1980, pp.401-420

Nelson, Leonard (ed.): Vier Briefe von Gauss und Weber an Fries. In: Abhandlungen der Fries'schen Schule (N.F.), 1 (1906), pp.431-440

Nový, Lubos: Origins of Modern Algebra. Leyden/Prag 1973

Rönne, Ludwig von: Die höheren Schulen und die Universitäten des Preußischen Staates. Berlin 1855

Scheibert, Carl G.: Die Prinzipien der Kombinationslehre nebst einer Bezeichnungsmethode in derselben. Programm des Gymnasiums Stettin. 1834

Schubring, Gert: On the relation of institutionalization of mathematics and professionalization of teacher education. A critical re-analysis of the plans of a polytechnical school in Berlin. Preliminary version, 1979 (Forthcoming in: Historical Studies in the Physical Sciences)

- : Bedingungen der Professionalisierung von Wissenschaft. Eine vergleichende Übersicht zu Frankreich und Preußen. In: Lendemains, no. 19, 1980, pp.125-135. (1980 a)

- : On Education as a Mediating Element between development and application: the Plans for the Berlin Polytechnical Institute (1817-1850). In: Jahnke/Otte (eds.) Epistemological and social Problems of the Sciences in the early 19th Century. Dordrecht 1980. (1980 b)

Sève, Lucien: La Philosophie Francaise Contemporaire et sa Genèse de 1789 à nos jours. Paris, 1962

Shinn, Terry: The French Science Faculty System, 1808-1914. Institutional Change and Research Potential in Mathematics and the Physical Sciences. In: Historical Studies in the Physical Sciences, 10 (1979), pp. 271-333

Stegmüller, Wolfgang: Theoriendynamik und logisches Verständnis. In: W. Diederich (ed.): Theorien der Wissenschaftsgeschichte. Frankfurt 1974, pp. 167-209

Stichweh, Rudolf: Ausdifferenzierung der Wissenschaft - Eine Analyse am deutschen Beispiel. Bielefeld, 1977

- : Differenzierung der Wissenschaft. In: Zeitschrift für Soziologie, 8 (1979), pp. 82-101

- : Disziplinen und Professionen. Unpubl. manuscript, 1980. Bielefeld

Tenorth, H.-E.: Professionen und Professionalisierung. Ein Bezugsrahmen zur historischen Analyse, in: M. Heinemann (ed. Der Lehrer und seine Organisation. Stuttgart. 1977

Thackray, Arnold and Merton, Robert K.: On Discipline Building: The Paradoxes of George Sarton. In: ISIS, 63 (1972), pp. 473-495

Tóth, Imre: Die nicht-euklidische Geometrie in der Phänomenologie des Geistes. Frankfurt 1972

Turner, R.S.: The Growth of Professorial Research in Prussia, 1818-1848, in: Historical Studies in the Physical Sciences 3(1971)

- : The Prussian universities and the research imperative. Ph. D. Thesis, Princeton

Hoëne de Wronski, Joseph M.: Introduction à la Philosophie des Mathèmatiques, et Technie de l'Algorithme. Paris 1811

CAMBRIDGE UNIVERSITY AND THE ADOPTION OF ANALYTICS IN EARLY NINETEENTH-CENTURY ENGLAND

Philip C. Enros

The first decades of the nineteenth century witnessed a great change in English mathematics. It was a time of revival marked by much distress over the state of mathematics in England and also by many efforts to rally from the English slump in mathematics of the eighteenth century. The period served as a threshold from the relative barrenness of the eighteenth century to the rich creations of such eminent English mathematicians as George Peacock (1791-1858), Augustus DeMorgan (1806-1871), and George Boole (1815-1864). Englishmen were once again to contribute to the mainstream of the development of mathematics.

This period in English mathematics has been portrayed, for the most part, as one of transition from the Newtonian dot notation and synthetic methods to the Continental differential notation and analytic methods [e.g. Ball, Becher, Cajori, Dubbey, Koppelman]. The description is barely sufficient. With its tacit assumption that a switch in mathematics caused the change, the account is of little help in understanding

---

This paper is a revised version of that given at the Berlin Workshop. It is based on sections of my doctoral dissertation "The Analytical Society: Mathematics at Cambridge in the Early Nineteenth Century", University of Toronto, 1979.

the timing of the revival, the motivations or intentions of the actors involved, or the direction in which mathematics in England was going. The renewal of English mathematics involved much more than simply a switch in notation and methods. Other important currents, some social in nature, played significant roles in the transformation which brought about the adoption and assimilation of Continental mathematics.

The University of Cambridge became an important center for the English adoption of foreign mathematics in the early nineteenth century. Its history provides a good example of the diversity of factors which were involved in the revival of mathematics in England. The aim of this paper is to show the chief ways in which Cambridge, as an institution, acted in the process of revival. Before examining the University's role, it is necessary to outline part of the background to the events. Two factors are especially important for understanding the developments at Cambridge: contemporary opinion about the state of mathematics in England, and the position of mathematics at Cambridge.

By the end of the eighteenth century many persons in Britain began looking to the Continent, and in particular to France, for advanced knowledge in mathematics. Among these were the mathematicians John Playfair (1748-1819) and Robert Woodhouse (1773-1827). They lamented the decline , or stagnation, of British mathematics. One of the causes of inferiority, they felt, was the traditional British stress on synthetic mathematics to the neglect of analytics.

"Analytic" denoted a particular style of mathematics. It had come into fashion in mathematics on the Continent in the second half of the eighteenth century largely through the works of L. Euler (1707-1783) and J.L. Lagrange (1736-1813). Its main characteristic was the formal manipulation of equations, or expressions; analytics implied an algebraic or formal, operational approach to a topic. The alternative style was synthetics. This was all that was not algebraic. During the latter half of the eighteenth century synthetics came to include all that was not strictly analytic. Hence the Newtonian style of the calculus, the theory of fluxions, was synthetic because it involved the idea of motion, a concept which was held as not algebraic. With the great achievements in mathematics and in mathematical science on the Continent at this time, non-analytic methods came to be identified with British mathematical inferiority. The adoption of analytics with its related differential notation, therefore, was seen by many in England as a remedy for the stagnancy of mathematics there.

There was, however, one other cause of the decline which was frequently mentioned, particularly by those who were in favor of change: the lack of public institutional encouragement for the mathematical sciences [e.g. Toplis, Thomson]. John Playfair of the University of Edinburgh, for instance, was not alone in arguing that the true cause of English inferiority lay in the state of the English universities and of the Royal Society [Playfair 1808]. The latter, for example, did not offer "sufficient encouragement for mathematical learning", unlike the Paris Royal Academy of Sciences which promoted mathematics by "small pensions and great honours, bestowed on a few men for devoting themselves exclusively to works of invention and discovery." [Playfair 1810, 398] Furthermore, Playfair held the English inadequacy in the mathematical sciences to be a result of the English public's self-defeating "mercantile prejudices" which were always prepared to demand an immediate justification for science in terms of use.

Playfair's lament was typical of the opinions of many others in Britain interested in mathematics. It was not just a reflection of the backwardness of English mathematics, but also a signal of a change in attitude in England towards mathematics. One of the key problems in understanding the development of English mathematics in the early nineteenth century is to provide a satisfactory explanation of why there was a concern for the state of English mathematics at that time. A tentative solution, which will not be developed in this paper, is that the concern was a reflection of the progress of the professionalization of mathematics in England. In any case, the widespread nature of the lament revealed a movement to renew English mathematics that had linked a style of mathematics with the advancement of mathematics.

Mathematics occupied a very important place in the system of Cambridge studies. Indeed, its prominence was one of the two aspects of Cambridge which distinguished that University from others in Britain. The second feature was Cambridge's final examination, the Senate House examination, which later evolved into the Mathematical Tripos. Young men going up to Cambridge in the early decades of the nineteenth century would enter one of its seventeen colleges. These controlled to a large extent the instruction of students. Besides classics, college lectures concentrated on mathematics, thereby maintaining a Cambridge tradition of mathematical study [Winstanley]. The lectures, especially in the larger colleges, covered such mathematical topics as Euclid, algebra, conic sections, plane and spherical trigonometry, statics, dynamics, hydrostatics, plane astronomy, fluxions, fluents, and Book I of the _Principia_ [compare Airy, Schneider, Wright, and Academicus]. A fairly good basic training in mathematics was available to, and expected of, almost all Cambridge students enrolled for a Bachelor of Arts degree.

While the average Cambridge student probably acquired a rather low level of proficiency in mathematics, quite a few did much more than was demanded by the college lectures [Airy, Wright]. Part of the reason for this was the few formal requirements of a Cambridge education as well as the inclination of some students towards mathematics. But the main motivation was undoubtedly the University examination, the Senate House examination, most of which was devoted to mathematics. The examination was held at the end of the period of study for the Bachelor of Arts degree, about 3 1/3 years, and was by far the most important and most rigorous test in qualifying for that degree. Although a very little knowledge might suffice for passing in the early nineteenth century, there was no maximum for the competition to be a wrangler, that is, to be in the first class of the honours list. Serious students, or "hard reading men", soon outstripped the college lectures by private tuition and study. This meant the working of problems in such periodicals as Thomas Leybourn's _Mathematical Repository_, the study of advanced topics such as the remaining books of the _Principia_, and increasingly in the early nineteenth century the effort to study such French mathematical works as those by S.-F. Lacroix (1765-1843), P.S. de Laplace (1749-1827) and J.L. Lagrange [Airy, Wright]. Besides fame and glory the reward for the Cambridge wrangler almost certainly included a valuable college fellowship, an important career consideration especially for those with few prospects [Tanner]. The Senate House examination, therefore, served as an institutionalized incentive for the study of quite high-level mathematics.

The University of Cambridge, like many other British institutions of that time, was confronted by a general spirit of reform.

> ...just as the University in the eighteenth century reflected the dislike of that age to violent change, so in the nineteenth century it responded to the prevailing sentiment that institutions, however venerable, had duties to the present as well as obligations to the past. [Winstanley, 157]

Much criticism from both within and outside of Cambridge was directed at the curriculum. There were some efforts, mostly ending in failure, to make the course of studies more comprehensive [Winstanley, 66-68, 167; Roach, 221]. Despite these attempts, mathematics continued to enjoy its privileged position in the intellectual life of the University especially in the acquiring of honors. This situation could not fail to be coupled with the lament about the state of English mathematics, particularly in that age of reform. The synthetic mathematics studied at Cambridge, the few alumni who pursued mathematical research, and the superficial stimulus to learning provided by the Senate House examination were all pointed to as proof of English stagnation [*e.g.* Playfair or Brougham 1816]. Indeed, the state of affairs at Cambridge was to rouse a number of individuals to attempt reforms in the mathematical studies. And the agents of change were mostly to be found among the students, not among the fellows.

Students were coming to Cambridge in the early nineteenth century, according to Sheldon Rothblatt, in a questioning mood. They were more independent than students of the eighteenth century and were "introducing into their university lives many of the social and intellectual ideas of their time" [Rothblatt 1974, 301-303]. It is then, perhaps, less surprizing, given the turmoil of this period of British history, that there was much dissatisfaction among students with the content and system of Cambridge studies. In particular, many students, reflecting the widespread regret about British mathematics, were unhappy with the synthetic mathematics of Cambridge. The structure of Cambridge was to foster this dissatisfaction.

> Students at our universities, fettered by no prejudices, entangled by no habits, and excited by the ardour and emulation of youth, had heard of the existence of masses of knowledge, from which they were debarred by the mere accident of position. They required no more. The *prestige* which magnifies what is unknown, and the attraction inherent in what is forbidden, coincided in their impulse.
>
> [Herschel 1832, 545]

The best example of a product of the forces mentioned above is the Analytical Society (1812-1813). It was a short-lived association of a small but remarkable group of Cambridge students, including John Herschel (1792-1871), Charles Babbage (1791-1871), and George Peacock (1791-1858). The Society was one of a large number of student associations at Cambridge of varying degrees of formality and size. But its aim was a reflection of the concern for the inferiority of English mathematics. Prompted by a familiarity or a proficiency or simply an enthusiasm for Continental mathematics, as well as by the widespread lament about the decline of English mathematics and by a dissatisfaction with the system and content of Cambridge mathematical studies, a number of students and one recent graduate decided to organize themselves. They resolved to contribute to English mathematical science by studying and advancing analytics. The members pursued this goal by electing officers, renting a room, starting a library, holding regular meetings, reading papers, and by publishing some of their research, the *Memoirs of the Analytical Society, for the year 1813*. The Analytical Society saw itself as a mathematical organization participating in the revival of English mathematics by the creation of analytical mathematics.

While Cambridge of the early nineteenth century could act as a stimulus to revival movements in mathematics, it also certainly was an obstacle to the study of analytics. Once again, the history of the Analytical Society provides a good example. The members, nearly all of whom graduated as high wranglers, were very concerned with the hurdle of the Senate House examination. Preparation for the examination limited the time that could be spent on the Society's activities. Furthermore, the Society's pursuits had no bearing on the degree of success in the Senate House because of the stress on synthetics there. By early 1814 nearly all of its members had graduated. They left Cambridge for various parts of England often to pursue careers which did not involve mathematics. The Analytical Society, therefore, was very much a child of Cambridge.

An informal mathematical revival movement emerged at Cambridge in the 1810s out of the dissatisfaction and feelings of deficiency that prevailed there, particularly among certain students and recent graduates. Cambridge not only played a role as a stimulus to this activity, but also molded its efforts. The University, through the structure of its studies, was to influence the ways in which the attempts to "reform" Cambridge mathematics expressed themselves.

Analytic mathematics found its way into Cambridge teaching very early in the 1810s. Many recent graduates used the customary wranglers' practice of private tuition, made possible (as noted above) by the meagre college teaching at Cambridge, to diffuse their "true faith" of analytics. George Peacock, John Herschel, Richard Gwatkin (1791- ?), and John Whittaker (1790-1854), all taught their private pupils French mathematics. Some graduates also went on to direct students' studies to "better" mathematics through the position of college tutor or lecturer. William Whewell (1794-1866), for example, became assistant tutor and mathematical lecturer at Trinity College in 1818. Eager to promote analytics at this time, he saw his new office as an opportunity to change the mathematics taught at Trinity [Todhunter 2 1876, 30].

The Senate House examination was another aspect of the University which was important for the adoption of analytics. Although the influence of the examination on the content of Cambridge studies had served as an obstacle to the introduction of analytics, the fact that the examination could exercise such an influence provided a means for altering those studies. George Peacock, appointed a Moderator of the examination in 1817, attempted to use his position to make changes in both its content and conduct [Peacock 1816, 1817]. His efforts at that time largely failed. However, he was once again Moderator in 1819 and this time, with the support of the other Moderator, Richard Gwatkin, and of one of the Examiners, Fearon Fallows (1789-1831), he was more successful [Peacock 1818, 1819]. The emphasis of 1819 was maintained in the 1820s by the Moderators, most of whom were also college tutors and it appears, committed to analytics. The control of the Senate House examination was a very important element of the successful diffusion of analytical mathematics in the Cambridge course of study [Herschel 1832, 545; Tanner].

Cambridge textbooks were of equal importance with teaching and the Senate House examination as a vehicle for change. The analytical movement supplanted, mainly in the 1820s, the old standard textbooks containing synthetic mathematics with new analytical ones or with translations of French works. Robert Woodhouse's (1773-1827) various texts from 1803 did much to introduce English readers to continental developments in various branches of mathematics. Similarly the translation of Lacroix's _Traité élémentaire_ in 1816 and the compilation of the _A Collection of Examples_ (1820) by Babbage, Herschel and Peacock, were written to help replace synthetics in the elementary course at Cambridge. The goal was to revise the course of study with a concentration on pure mathematics and with an eye to keeping pace with the general advancement of the field. Many such analytical treatises appeared in the 1820s. As many of the authors were also Moderators, the contents of those works quickly found their way into the Senate House examination [Great Britain, 454].

Analytical mathematics was adopted at Cambridge very quickly in the late 1810s and early 1820s due to activities within the University of a new generation whose goal was to revive Cambridge mathematics. The

University had served as both a stimulus and a vehicle for that revival, but it was not merely a passive factor. The goals of the revival movement at Cambridge had parallelled the concern over the condition of mathematics in England. This anxiety, as noted above, involved not only a style of mathematics (analytics) but also a promotion of research in mathematics. The latter was also part of the outlook of the Cambridge movement. Herschel, for example, wished that the University would develop the student's "relish for mathematical speculation" and that it would encourage postgraduate studies in mathematics [Herschel 1816]. But his views were nothing more than wishful reflections of his own motivation. The research ideal did not fare as well at Cambridge as had the transition to analytics. Cambridge did not exist to promote mathematics and would not move in that direction, at least in the first half of the nineteenth century. The educational goal of Cambridge, a liberal education, was to temper the original impulses of the revival movement.

Mathematics at Cambridge found its meaning in education, in the ideal of a liberal education [Rothblatt 1976; McPherson 1959]. Both the content and the system of studies were justified by this ideal. It implied the molding of the character of a young man into that of a gentleman. Such an education stressed the transmission of the culture of man or of the nation to the individual. A liberal education existed in sharp contrast to any education devoted solely to specialized training for a later career. While Cambridge students might receive a very good training in mathematics, the purpose of the University was not to train mathematicians nor to push back the frontiers of mathematics.

An appeal to the idea of a liberal education could have implications for analytic mathematics as well as for research. An important facet of the distinction between analytics and synthetics was a commonly held opinion about the difference in their value. Probably due to the great advances in mathematics of the seventeenth and eighteenth centuries, analytics was highly regarded for its power of discovery. It was the best example of the way in which reasoning was to be used [see "Analytics" in, for example, Hutton 1795 or Barlow 1814]. Analytics was therefore firmly linked with research mathematics in early nineteenth-century British thought. By contrast, synthetics was prized for the clarity and rigor of its explanations. Many persons had misgivings about the vagueness and imprecision then associated with analytics. Synthetic mathematics was linked to education because of its aptitude, which had been traditionally acknowledged, for developing and strengthening the reasoning powers of the mind. A liberal education, in itself, would therefore favor synthetics in the curriculum at the expense of analytics.

Analytics at Cambridge did meet with some such criticism although the censure did not prove to be strong enough to prevent the adoption of the new mathematics. The _London Magazine_ saw the triumph of analytics over geometry as "one more proof how strongly the tide of opinion at Cambridge sets in towards the belief, that men are congregated in those Boeotian flats for the promotion of science, rather than of education" [Anon, 303]. Similarly, Arthur Browne, a fellow of St. John's College, argued that any superiority which analytics had over geometry was valuable only to those who intended to devote their whole lives to mathematics. But the object of a university, he thought, was not to expand science but to diffuse religious knowledge and to supply men qualified for offices in the Church and in the State [Browne, xiv-xv, xviii-xx].

Analytics, as had been the promotion of research, was to be rejected by the circumstances of Cambridge and frustrated by the ideal of a liberal education. It seems that increasing criticism of Cambridge, particularly in the 1830s and 1840s, gave rise to a defensive reaction within the University. This response manifested itself in mathematical studies by an emphasis in the curriculum on geometry and elementary mathematics and by an assertion of the subservience of mathematics to the goals of intellectual discipline. Henry P. Hamilton (1794-1880), for instance, abandoned the reliance on analytics and the stress on advanced mathematics in the fourth edition (1838) of his textbook *An Analytical System of Conic Sections* because they were "too scientific" [iii]. By 1850 Whewell was able to rejoice in the successful checking of the "mischievous tendency" of analytics [Great Britain, 500]. Thus the ideology of a Cambridge education was finally to triumph over analytics as it had over research.

The story of the adoption of analytics at Cambridge in the early nineteenth century illustrates the importance of context for understanding developments in English mathematics. In this case, the University was itself an important element of the process by which certain aspects of Continental mathematics were transmitted and assimilated. Cambridge was involved both through its system of studies and through its ideology. These served to mediate the movement which aimed at reviving English mathematics. The type of mathematics, as well as the role of mathematics, which was accepted at Cambridge University underlines the way in which any institution reflects society. Mathematics was used to educate gentlemen, not to train mathematicians. It had not yet been accepted as a profession in early nineteenth-century England.

References

Academicus (1801) "A letter on the 'Course of Studies at Cambridge and Senate-House Exam'" *Monthly Magazine* 11, 115-118, 292-294.

Airy, Wilfrid (ed.) (1896) *Autobiography of Sir George Biddell Airy*. Cambridge (University Press).

Anon (1826) "The Cambridge University" London Magazine 4, 289-314.

Ball, Walter W.R. (1889) A History of the Study of Mathematics at Cambridge. Cambridge (University Press).

Barlow, Peter (1814) A New Mathematical and Philosophical Dictionary. London (G. & S. Robinson).

Becher, Harvey (1971) "William Whewell and Cambridge Mathematics" Ph. D. dissertation, University of Missouri, Columbia.

Browne, Arthur (1824) A Short View of the First Principles of the Differential Calculus. Cambridge (J. Deighton & Sons).

Cajori, Florian (1919) A History of the Conceptions of Limits and Fluxions in Great Britain from Newton to Woodhouse. Chicago (Open Court).

Dubbey, John M. (1978) The Mathematical Work of Charles Babbage. Cambridge (University Press).

Great Britain (1852) Cambridge University Commission. Report of Her Majesty's Commissioners &c. Parliamentary Papers, Command No. 1559.

Herschel, John (1816) "Review of Dealtry's The Principles of Fluxions" 19pp. unpublished: Herschel Collection, Humanities Research Center, University of Texas.

(1832) "Mrs. Somerville's Mechanism of the Heavens" Quarterly Review 47, 537-559.

Hutton, Charles (1795-1796) A Mathematical and Philosophical Dictionary. London (J. Davis).

Koppelman, Elaine (1971) "The Calculus of Operations and the Rise of Abstract Algebra" Archive for History of Exact Sciences 8, 155-242.

McPherson, R.G. (1959) Theory of Higher Education in Nineteenth-Century England. No. 4, University of Georgia Monographs.

Peacock, George (1816, Dec.3) "Letter to Herschel" in Collection of Herschel Manuscripts in the Royal Society, London.

(1817, Mar.4) "Letter to Herschel" same location as 1816.

(1818, Mar.7) "Letter to Herschel" same location as 1816.

(1819, Jan.13) "Letter to Herschel" same location as 1816.

Playfair, John (1808) "Review of LaPlace's Traité de Méchanique Céleste" Edinburgh Review 11, 249-284.

(1810) "Review of Laplace's The System of the World as translated by John Pond" Edinburgh Review 15, 396-417

Playfair, John or Brougham, Henry P. (1816) "Review of Dealtry's Fluxions" Edinburgh Review 27, 87-98.

Roach, John P.C. (1959) "The Age of Reforms, 1800-82" A History of the County of Cambridge and the Isle of Ely, 235-265. Volume 3 of The Victoria History of the Counties of England. Oxford (University Press).

Rothblatt, Sheldon (1974) "The Student Sub-culture and the Examination System in Early 19th Century Oxbridge" in L. Stone The University in Society 1, 247-303. Princeton (University Press).

(1976) Tradition and Change in English Liberal Education. London (Faber & Faber).

Schneider, Ben R. (1957) Wordsworth's Cambridge Education. Cambridge (University Press).

Tanner, Joseph R. (ed.) (1917) The Historical Register of the University of Cambridge. Cambridge (University Press).

Thomson, T. (1815) "Review of Wainewright's Literary and Scientific Pursuits &c." Annals of Philosophy, 294-304.

Todhunter, Isaac (1876) William Whewell, D.D. 2 vols. London (Macmillan).

Toplis, John (1805) "On the Decline of Mathematical Studies, and the Sciences dependent upon them" Philosophical Magazine 20, 25-31.

Winstanley, Denys A. (1940) Early Victorian Cambridge. Cambridge (University Press).
Wright, John M.F. (1827) Alma Mater &c. London (Black).

A SURVEY OF FACTORS AFFECTING THE TEACHING OF MATHEMATICS OUTSIDE THE UNIVERSITIES IN BRITAIN IN THE NINETEENTH CENTURY

Leo Rogers

At the beginning of the nineteenth century, Britain was already well into its Industrial Revolution. When we consider that in the space of some hundred years the transport system alone developed from the use of riding track and canal, through coach road to a complex railway network and even the first motor cars and airships, we have an indication of the rate of the change and the ease with which ideas flowed along the communication network, ready to be utilised by those who saw their potential and advantage.

The roots of the nineteenth century changes in, and eventual institutionalisation of the means of education lie not only in the increase in industrialisation but also in the gradually accelerating view of the expectations of ordinary people.

Accompanying the industrial and social changes were deep changes in attitude towards science - particularly applied science and technology - which showed science as a benefactor and which provided the patronage for pure science to flourish.

The encouragement of the study of science by the rising middle class led to the establishment of a number of 'Literary and Philosophical' Societies, the most famous of these being, perhaps, the Lunar Society, who met originally at Mathew Boulton's Steam Engine Works in Birmingham from 1766 for discussion and experiment on immediate industrial problems and whose later interests developed into the advancement of science and technology and the area of social and political education. Many famous names are associated with this society, among them Watt, Keir, Galton, Priestly, Erasmus Darwin, Edgeworth and Wedgewood, who deliberately educated their children to be leaders of nineteenth century industry. The middle-class, forward-looking industrialist was also often a scientist-innovator as well, with a strong interest in practical applications.

Men like these came largely from non-conformist backgrounds. They had been excluded from any Public Office and from taking degrees at Oxford or Cambridge by their refusal to take the oath of allegiance to the King as head of the Church of England and so went to Scottish Universities or one of the Dissenting Academies. They were often Rationalists, attracted to French philosophies of the enlightenment, with their social and educational implications, supporting struggles for liberty and exposing corruption. Their main contribution to Education was to reject the values of eighteenth century aristocratic society by adopting versions of materialist philosophy and psychology and attempting to put forward new and relevant designs for social living.

For example, the curriculum at Warrington Academy (1757-1786) which set the model for many of the nineteenth century Academies and Colleges to follow favoured scientific enquiry, was open to the influence of the new needs of society, and included such revolutionary subjects as history, politics, modern languages, commerce and the practical applications of mathematics and science. The aim of such a curriculum was to prepare young men for their future role in the development of science and industry, and the people educated at Warrington and other places became the members of the numerous Literary and Philosophical Societies that flourished in the nineteenth century and who directed a decisive stage in the Industrial Revolution.

All the major centres of industry supported a society devoted to the furtherance of useful arts for the improvement of local industry. A typical example is the Manchester Literary and Philosophical Society. Founded in 1781, it supported the Manchester College of Arts and Sciences (1783) for part-time students where chemistry and mechanics were taught and considered to be most relevant to local industry. Other subjects already mentioned were included in the curriculum and also classical languages, grammar and rhetoric, mathematics (including trigonometry) and commercial and economic geography. Manchester Academy (1786) was soon founded to cater for full-time students, and both these institutions evolved later into Owens College (1853) and eventually into Manchester University. The Academies were supported by contributions from local industry, since the pay off - applicable knowledge - was direct and obvious, and the long-established traditions of craft-apprenticeship were fostered in the new technologies. Better communications also assisted a number of serials to flourish; the well-known Ladies Diary being one, and the readership of these and of technical and mathematical articles and even regular columns in newspapers, widened. The range of knowledge shown by correspondents to such columns was quite wide; as well as the expected algebra, arithmetic and trigonometry, we find, for example, mensuration, statics, dynamics, probability, calculus and conic sections. Mathematical columns in newspapers were generally short-lived, starting as recreations for interested 'philomaths' but often becoming academic and specialised, for the enthusiast only.

For the most part, the contributors were either self-taught in the sense that they had the leisure and access to the well-established 'non-university' mathematics of writers like Bonnycastle and Hutton and may well have had some training from one of the mathematical practitioners still thriving at the beginning of the nineteenth century; or they were already products of the Academies and Colleges exercising their newly acquired practical knowledge.

The rise of technology required greater servicing from the reviving University mathematics, and a number of interpreters and educators flourished who transmitted portions of this university mathematics to the common man. Such people as Olinthus Gregory, and Thomas Tate were making mathematics more available, and Augustus De Morgan's work was serialised by the Society for the Diffusion of Useful Knowledge and apparently read as avidly as the serialised work of his contemporary, Charles Dickens.

The work of Dickens and others prompted the social conscience of the middle class and was one of the factors responsible for the institutionalisation of social services and the gradual taking over of this responsibility by the state. Radical philosophy thus slowly helped to distribute the wealth created by the new technology, so that by the end of the nineteenth century universal education was a reality.

As indicated, the mathematics taught and used was essentially practical, as the samples below show - the criteria being the essential applicability to current problems in commerce, measurement of various kinds, and industry. Thus, subsumed under the general heading of practical mathematics may be a wide variety of skills - from instrument making to determining the density of a chemical solution for tanning hides.

As the institutions developed, and the subjects studied themselves became more complex, the necessary applications of mathematics were gradually taken in and taught within a particular subject area, to become part of 'physical chemistry' or 'mechanical engineering', etc., so that very often what once started as a good example of 'theoretical' mathematics applied to a real problem, became an isolated and possibly archaic rule of thumb. The applications of mathematics became too numerous and too specialised to be taught by mathematicians.

The public discussion and dissemination of mathematics in the nineteenth century helped to breed a new class of teachers, socially aware and devoted both to the teaching of mathematics and the educational development of children and young people. While the content of the elementary school curriculum was largely arithmetic and that of the grammar and public schools included some Euclid and trigonometry, this state of affairs had taken a century to come about, and was considerably accelerated from 1863 onwards by a Royal Commission enquiry into the curriculum of the public schools, and a number of subsequent Government Reports. Teachers were being asked their opinions, and in particular their voice was being raised against traditionalist attitudes on the content of the curriculum. The Association for the Improvement of Geometrical Teaching was formed, as a grass-roots 'anti Euclid' movement, but not so much for banning Euclid as for making the teaching of geometry more appropriate to the mental development of children. From this beginning the Mathematical Association was to develop the four-stage plan for the teaching of geometry, from practical drawing, to proof.

The evolution of the mathematics curriculum in the nineteenth century is a highly complex affair. This brief summary suggests that the major influence as to content came from the developing needs of industry, and that of method from the growing social awareness and general level of expectation of the teachers involved. How these two interacted and exactly what influence British University mathematics had on the schools, must be subject for continued examination.

SAMPLES FROM THE NINETEENTH CENTURY MATHEMATICS CURRICULUM

(i) Dilworth, Thos. The Schoolmasters Assistant. Being a Compendium both Practical and Theoretical., Eleventh edition, London 1780, (Various editions from c. 1740).

Typical of most basic arithmetic teaching throughout the nineteenth century both in content and implied methods.

Contents: 1. Arithmetic in whole numbers and the common rules, (addition, subtraction, multiplication, division). This simple heading contains a very large number of examples of all kinds of 'applications' of arithmetic, each categorised as a particular skill or technique.

The start of each section is a kind of 'catechism' or series of questions and answers (which the pupil was clearly required to learn by heart) which defined the skill about to be learnt and stressed its relevance in theory or practice.

For example, from the Introduction (pp 1-2).

"Q. What is Arithmetic?

A. Arithmetic is the art or science of computing by Numbers, either Whole or in Fractions. - - - - - - -

Q. What is Theoretical Arithmetic?

A. Theoretical Arithmetic considers the Nature and Quality of Numbers and demonstrates the Reason of Practical Operations. And in this Sense Arithmetic is a Science.

Q. What is Practical Arithmetic?

A. Practical Arithmetic is that which shews the Method of working by Numbers, so as may be most useful and expeditious for Business. And in this Sense Arithmetic is an Art.

- - - - - - - - - - -

Q. Which are the Fundamental Rules in Arithmetic?

A. These five: Notation, Addition, Subtraction, Multiplication, and Division.

Q. What is Notation?

A. Notation is the Art of Expressing Numbers by certain Characters or Figures."

A curious reason is given for there being nine digits or nine places, rather than eight or ten; namely that a number of nine digits (say 123,456,789) was sufficiently large "to express most ordinary concerns" (p2).

Addition is simple, that is the addition of numbers representing the same type of object, pounds to pounds, yards to yards , etc. and Compound, that is the addition of mixed quantities, like shillings and pence. Similarly, we have simple and compound subtraction, multiplication and division. Note that because this section deals only with the arithmetic of whole numbers, any remainder of a division is not expressed as a fraction. Division "shews how oft one number is contained in another, and what remains". (p.31).

These remarks indicate what we now would regard as a very complex system in common use, where it is possible to conceive that a person may be able to operate with simple calculations in his own trade, but not be able to transfer his skill to another, because they had been set up by teachers as conceptually distinct. I have no firm evidence for this; it is a speculation from a modern point of view and must be regarded in the appropriate cultural context.

The applications are given in many detailed examples of various trades, where the quantities and measurements differ widely. Almost every trade had its peculiar unit of measure, some of which are clearly local, while others were regarded as rather special applications, For example, "the Denominations of Motion in the heavenly Bodies" was angular measure. (p.20).

12 signs or 360 Degrees make the circle of the Zodiac

30 Degrees make 1 sign

60 Minutes make 1 Degree

60 Seconds make 1 Minute

Division subsumed reduction, (both descending and ascending) the rule of three, direct and inverse proportion, and practice. In fact, this section demonstrates the applications of multiplication and division in the calculations of bills, quantities, etc. A great deal of attention is paid to the setting out of calculations, as would be expected in book-keeping and examples are given of interest, discount, equation of payments, exchange, alligation, arithmetic and geometric progression and permutation.

This accounts for a good half of the contents and where the units of measure are so varied and contain so many sub-units, it seems unlikely in practice that traders ever required to manipulate fractions extensively. As today, they no doubt had traditional methods of allowing for left-overs or small deficits.

In this sense, the arithmetic taught here was much more relevant to actual practice, than the arithmetic taught in our schools today. Pupils knew if they got it wrong, they starved. The other sections of the book contain:

2. Vulgar Fractions: Notation, Reduction and the four Rules of Three, Direct and Inverse. (A short section of only ten pages).
3. Decimal Fractions: Content as above with vulgar fractions and then sections on powers and roots, interest, annuities, rebate and the equation of payments.
4. The next section contains a large selection of questions (and answers) for practice.

5. The last section (of twelve pages) is a rather specialised piece on "Duodecimals" where the examples given are of a joiner measuring wood in feet, inches, seconds (twelfths of an inch), and so on. A short table giving decimal equivalents to duodecimals is displayed and some examples of its use given.

This last may seem a curiosity, but two aspects are significant. In the first place, it represents a consistent base-twelve system:

"12 Fourths make 1 Third
12 Thirds make 1 Second
12 Seconds make 1 Inch
12 Inches make 1 Foot" (p.181)

and one could imagine that the system was chosen (or built up purposely) to display convenient combinations of halves and thirds. Of course, it is unlikely that carpenters ever needed to measure to an accuracy of more than a second (twelfth), and in actual practice most measurements are not taken with a ruler, anyway.

Secondly, it is the only example in the book where a table is used to demonstrate and calculate equivalences, in the manner of a ready-reckoner.

As usual, in such a book, we have a fairly lengthy introduction, containing an essay on the education of youth, notable for its plea for the education of girls; for the woman who has had a liberal education "--knows the Advantages that arise from a ready Use of the Pen ---" (p.xiv) which give her some independence, particularly in widowhood. There is also a list of some fifty-two Lecturers, School-masters, Writing-masters and Teachers, the majority from near London, who recommend this book as the best for the "--speedy improvement of YOUTH IN ARITHMETIC--" (pp xix-xx).

(ii) Crossley, J.T. and Martin, W.

The Intellectual Calculator, or Manual of Practical Arithmetic --- London. 58th Edition. c.1870.

The contents, and even the examples are virtually the same as Dilworth, while the "Duodecimals" appear, though not by name, as part of an exercise of the "Fractograph" (a geometrical diagram for demonstrating simple fractions) on the last two pages (145-146) of the book.

Gone are the exhortations, recommendations and the detailed examples of setting out; much is condensed into lists and tables so that "... the rationale of the science should be demonstrated in a manner calculated to draw forth the thinking powers of a child." The authors have "... kept in view the market, the counting-house, and the shop." (p.7) The book also includes a "complete course of mental arithmetic" reduced to a regular system which the student was supposed to be able to apply to any situation.

Since the book was so popular, it would appear that memory was a major criterion in the learning of arithmetic at this time.

(iii) Davidson, J.

A System of Practical Mathematics: Edinburgh: 1832

(Third Edition)

Davidson's text was intended for use in Public Schools (i.e. Private Schools and Academies where more advanced and specialised mathematics was taught) and again is fairly typical of the style of nineteenth century teaching.

I let the following extracts from Davidson's preface and contents speak for himself.

(iii - xv): ".. it is still the general opinion of experienced teachers, that a plain treatise, comprehending the best practical rules, with examples of their use and application, accompanied, where necessary, with explanatory notes and practical remarks, a sufficient number of well-selected and accurately expressed unsolved exercises adapted to each rule, and demonstrations so elementary as to be intelligible to anyone who understands Arithmetic and the Elements of Geometry and Algebra, is much wanted: and such a course of instruction I have endeavoured, to the best of my ability to supply ..."

The contents are as follows:

I. Algebra. This deals with all the rules and processes we have previously seen in books for the elementary school (excluding weights and measures) as 'generalised arithmetic' to display the 'first principles'.

II. Elements of Geometry. Contains some theorems of plane Geometry with emphasis on proportion and construction, with geometrical problems ".. by drawing which the learner will acquire the use of his instruments ...". In this section also we find procedures for determining the accuracy of "rhumbs, chords, sines, tangents and secants" and for finding missing values in tables.

III. Plane Trigonometry includes the construction of trigonometrical tables and the construction and use of 'Gunter's scale' (a version of the slide rule for use on trigonometrical calculations) and the 'sliding rule' (a specialised slide rule for estimating quantities and prices in the measurement of timber and other bulk products).

IV. Mensuration of Heights and Distances includes "the distance of an object by the motion of sound," (e.g. the time between sighting an explosition and hearing it), heights and depths using the barometer and the motion of heavy bodies.

This is followed by -

V. Mensuration of Surfaces

and a chapter on -

VI. Conic Sections: the parabola, ellipse and hyperbola.

next comes -

VII. Mensuration of Solids with a large number of examples of all kinds of sections of various solids.

This concludes the more general part of the text (about a third) and the following sections deal in greater detail with a wide variety of practical skills, namely:

VIII. Specific Gravity (including weight of cattle)

IX. Land Measuring (including the use of plane table and theodolite)

X. Artificers Measuring (timber, brickwork, etc.)

XI. Gauging (including residues left in casks)

XII. Gunnery (including the use of triangular numbers to count piles of balls or shells).

XIII. Spherical Trigonometry

XIV. Geographical and Astronomical Problems

XV. Methods of Ascertaining Time

XVI. Methods of finding the Lattitude

XVII. Methods of finding the Longitude and finally,

XVIII. Navigation (both plane and globular)

The last section of the book (some 200 pages) is taken up with various tables of weights and measures, and a range of logarithmic, trigonometrical and astronomical tables, together with methods for their calculation. These tables "... were collated with the tables of Hutton, Callet, Taylor, Briggs, etc., by which means, many errors, especially in the last decimal figures, which had escaped former editors, were discovered and corrected."

Given the comprehensive nature of this text, and the straight forward practical attitude of its author, it is no surprise to see a note of guidance for the reader: "... that though the different parts of the work are so arranged that they may properly be taken in succession, yet they are in general so distinct, that a learner who has not leisure to go over the whole, may select and study any particular branch, which is more the object of his pursuit than the rest".

These selections from the 'mathematics of the common man' of the nineteenth century show a marked contrast with the academic university studies of the same period. We know the state of university mathematics in England at the beginning of the century was poor indeed, and by the end of the century a considerable amount of 'catching up' had been done, with mathematicians from the British Isles making some outstanding contributions to the general body of knowledge. During the century we also have attempts at popularisation, for example by De Morgan in the Penny Cyclopaedia - a weekly pamphlet which serialised his Calculus and other expository mathematical writing from 1837.

In spite of the growing influence of England in mathematics, and the attempts at popularisation, very little change occurred in the content of the mathematics taught in schools, colleges and academies. Elementary schools taught arithmetic, the Public Schools set the model for the secondary sector with Euclid, Algebra and Trigonometry for those who aspired to university entrance, and the colleges and academies taught the specialised and fundamentally practical mathematics required for commerce and technology.

3. General References

Hobsbawn E.J. (1968) Industry and Empire London.

Simon B. (1974) The Two Nations and the Educational Structure London.

Society for the Diffusion of Useful Knowledge (1833) The Penny Cyclopaedia (1833-1846) London.

Report of the Royal Commission appointed to enquire into Revenues and Management of Certain Colleges and Schools (Clarendon Report) (1863) London.

Report of the Royal Commission on Scientific Instruction and the Advancement of Science (Devonshire Report) (1872) London.

MATHEMATICS IN A UNIFIED ITALY

Umberto Bottazzini

1. "The Risorgimento, the national rebirth of Italy, also meant the rebirth of Italian mathematics" (Struik, 1967,179) or, in Volterra's words, "the scientific existence of a nation" (Volterra, 1900,43). An idea of the difficulties - political and otherwise - met by Italian mathematicians before the unification of Italy may be derived from the fact that Mossotti, physicist and mathematician, and Betti's mentor in Pisa, upon his return to Italy after a long exile in England and the Argentie, was refused a chair in the Lombardo-Veneto and in the Papal State because of his patriotic ideas. Another instance: travelling from Austrian-ruled Pavia, where Brioschi was teaching, to Rome in the Papal State, where Tortolini's _Annali di Scienze Matematiche e Fisiche_ were published,

required crossing no less than four state borders. This situation made the direct exchange of ideas among mathematicians from the various states very difficult, or altogether impossible. Not even the few meeting of the Italian Scientific Society (called "Società dei XL") were of any help; in fact, this association was looked upon suspiciously by the police of the various states and gave a number of informers, probably present even among the mathematicians, the opportunity to report on the conspiratorial and patriotic activity of the scientists. Little wonder then that only in 1858 men such as Brioschi and Betti, Tardy and Genocchi were able to meet in person - personal contacts being so meaningful for mathematical development. This meeting occurred on the occasion of the founding of the new Annali di Matematica Pura e Applicata modelled after the German (Crelle's Journal) and the French (Liouville's Journal). Nor does it appear to be a casual circumstance that this new journal first appeared in 1859, when the movement towards the political unification of Italy started with the annexation of Lombardo-Veneto to Piedmont (the nucleus of the Italian nation, as Prussia was to the German) and with the first awakening of a national political conscience. Proof of the latter is found in the editors' opening statement in the first issue

of the *Annali*:
"The editors' trust (nor would they otherwise have undertaken this publication) that the Italian geometricians will do their utmost in order to make sure that a journal aiming to represent our science is able to continuously attract the attention of the learned of other countries , thus putting an end to the complaint that our work is unknown abroad" .

This trust was confirmed when the *Annali* - edited by Brioschi and Cremona from 1867 - grew to be one of the most authoritative European journals.

In the Risorgimento years the young Italian mathematicians took an active part in political life; first in the independence war and later in building a new country. "Several of the founders of modern mathematics in Italy participated in the struggles which liberated their country from Austria and unified it", wrote D.J.Struik (1967, 179) and, elsewhere: "The relationship between scientific selfconsciousness and the struggle for independence is personified in the young mathematicians Betti and Cremona, who were soldiers in the wars for political freedom"(Struik 1979). Actually, the former was a volunteer in the Pisa student battalion led by Mossotti, while the latter fought against Austrians in Venice. The Austrian police also kept their eyes on Casorati and Brioschi.

With the foundation of the new state, the situation, as concerns mathematics, changed thoroughly. Among the intellectuals who took part in the country's political leadership, after unification 1861, the scientists, and especially the mathematicians, played an important role. Betti's friend, the Pisan physicist Matteucci became made Minister of Education in 1862, with Brioschi as General Secretary. From the inception of the new state, Brioschi and Betti, first as members of parliament and later as senators, as many a mathematician later on: Beltrami, Cremona, Dini, Volterra, etc., were appointed to the High Council for Public Education. On the institutional level Casati's law (1859) favoured the establishment of new chairs: the first chair of Higher Geometry at Bologna University (1860) was given to Cremona, and teachers and student could now move freely from one university to another: for instance, Beltrami first taught in Pavia then in Bologna and Pisa. "The first, enlightened decision of the national government was the institution of special chairs for the teaching of higher mathematics, and the appointment to these chairs of the famous men whom we mentioned, who were then followed by other, no less famous personalities. Thus a new environment was suddendly formed and a new era began"

(Volterra 1908,58).
To the names mentioned so far Genocchi and Battaglini should be added, the former left his home-city, Piacenza, in 1848 before the Austrians returned victorious and, after settling in Turin, began teaching higher analysis at the university in 1860, while in the same year Battaglini began his course of higher geometry at the university in Naples. A significant indication of the changes brought about by national unification, for instance, is the fact, also recorded by Volterra, that "Battaglini was not entrusted with public teaching before 1860: he had failed an examination because he had dealt with the problem according to the new, fruitful ideas of Salmon instead of using Newton's older methods" (Volterra 1908,57) .

2. In connection with the most advanced currents of European research, the Italian mathematicians of this period are the protagonists of a strong, thorough renewal of the national mathematical tradition.
"These mathematicians, like Cavour in politics, turned their eyes toward Germany, a country that was emerging out of a maze of smaller and larger states into a strong empire with an equally strong mathematical establishment. They studied Gauss, Riemann, Clebsch and later Klein, without neglecting the French. By the time that Italy had a-

chieved its unity with Rome as its capital, it could be proud of a group of mathematicians, Beltrami, Betti, Brioschi, Codazzi, Cremona and others, with an international reputation. For mathematics the Risorgimento was a Rinascimento" (Struik, 1979).

This Rinascimento had, of course, roots in the mathematics of the early nineteenth century, although the historians have often forgotten them, emphasising the work of Betti, Brioschi, Cremona, Dini and so on. In the early decades of the past century a large group of italian mathematicians was working in mathematical physics: even if little known nowadays, they were then, as a group, a very important one. People such as Bordoni, Mossotti, Plana, Piola, Chiò and others were influential and esteemed at home and abroad. The true novelty was in the field of research of the young mathematicians: they abandoned the prevailing french tradition in applied mathematics and studied algebra and group theory, invariant theory, elliptic functions and complex analysis, projective and algebraic geometry.

Besides research, the outstanding Italian mathematicians also dedicated their energies to teaching both at secondary schools and university level. This in my opinion is an important factor for understanding the development of

Italian mathematics in the second half of the last century, when the formation of "schools" around the most prestigious teachers played a significant part.

Along with pure research, an effort was made to train qualified technicians and engineers, needed by the country for its industrial take off. Thus, in 1863, with the financial help of the Lombardy industrialists, Brioschi founded an engineering school in Milan (the present Polytechnic Institute) to turn out the engineers required for the industrial transformation of Northern Italy which chiefly needed an efficient railway network. Among the teachers at the engineering school there were Brioschi, Casorati and Cremona. A similar school was founded by Cremona in Rome.

At this time, the essential problem in teaching was the drawing up of adequate manuals and treatises, based on the latest research: this problem was enthusiastically tackled by the young mathematicians.

Cremona and Casorati were thinking of a treatise on higher algebra when they learned from Novi that the first volume of his own treatise (Novi, 1863) - based on the lectures delivered by Betti a few years earlier - was being printed. Novi, instead, suggested that they consider a manual on analytic geometry. He wrote to Casorati:"Did you never think about a treatise on analytical geometry? Tardy is working on his book on differential and integral calculus. Let's see if we can apply rigour to our studies in Italy". And, when the project seemed to be abandoned, Novi again exhorted them to carry on and added: "Thus, if I succeed in making Tardy publish his treatise on differential and integral calculus which has kept him busy for several years, we shall have a complete Italian course on higher mathematics". However, neither Tardy's work (Bottazzini 1980,85) nor the vaguely outlined book by Casorati and Cremona ever appeared. Casorati taught analysis at Pavia and wrote a treatise on complex analysis (Casorati, 1868) while Cremona issued, a few years later, a projective geometry handbook (Cremona, 1873).

However, the leading Italian mathematicians were very busy translating foreign books, treatises and papers; this tradition would continue without interruption until the end of the century.

Again, the scientific imprint given to the reform of secondary education (1867) was due to the mathematicians who were members of the governing bodies of the Ministry of Education; this reform concurred with the contemporary law on the "suppression of ecclesiastical bodies" (1867) and marked a turning point in the secularisation of the Italian state. In fact, the still open "Roman Question" - Rome, the Papal seat, was still severed from the rest of the nation - represented the main political problem of the newly born nation at the time. The attitude of the young Italian mathematicians is apparent from certain letters of theirs: faced with Tortolini's ambiguity concerning the issuing of the Annali di Matematica Pura e Applicata (1858) Brioschi did not hesitate to remind him and his Cardinals that the time of Galilei's trial was long past; while Casorati wrote to Betti in 1860 and wished Rome to be "freed of papal tyranny" as soon as possible; Cremona was similarly firm who invited Betti and Brioschi, members of the Higher Council for Education, to put and end to the shameful requirement that, to be admitted to the universities of the new Kingdom, students should take an examination on "the mysteries of the Catholic faith".

This anticlerical, scientific attitude was one of the basic elements of Italian culture in the second half of the 19th century; an attitude which would last until the 1920s when the Catholics were readmitted to political life and while, at the same time, philosophic idealism ruled Italian culture.

Beside teaching at university level, the Italian mathematicians also took great interest in the problem of secondary education. In fact, as Loria said: "au moment où l'Italie devenue enfin libre put jouir, d'un bout à l'autre, d'un gouvernement national, elle conservait encore dans son organisation scolaire des traces déplorables et évidentes de son séculaire servage. Dans l'ancien Piemont p. ex., certainement à cause de l'influence française, on préferait la méthode demi-arithmétique de Legendre aux rigoureux procédés géométriques d'Euclide: tandis que dans les provinces qui venaient de secouer le joug autrichien se trouvaient répandus des manuels écrits avec le seul but évident de spéculation commerciale" (Loria 1904, 595).

A few years earlier (1856) Betti translated Bertrand's handbook of elementary algebra and in 1861 with Brioschi edited a new edition of Euclid's Elements. From 1862 they both worked at establishing a company whose main aim, as Betti writes to Casorati, "is to supply textbooks for secondary schools and to spread useful knowledge". On the other hand, Cremona, a member of the ministry's teaching programmes committee (1867), suggested that the <u>classical</u> secondary schools adopt Euclid's text as a basis for teaching geometry. "Si cette mesure, que le gouvernement s'empressa d'adopter, peut paraître aujourd'hui un peu trop draconienne, lorsqu'on tient compte du but qu'elle se proposait (et qu'elle atteint en effet), c'est-à-dire d'extirper de nos écoles les mauvaises habitudes introduites par certains livres, elle doit être considérée comme un des actes du grand mathématicien qui le signalent à reconnaissance éternelle de ses concitoyens" (Loria 1904, 595).

At the time, however, this decision met with strong opposition. The Giornale di Matematiche, a new journal "for the use of Italian university students" founded in 1863 by Battaglini, a staunch supporter of non-Euclidean theories, published in 1868 an anonymous translation of a speech delivered by J.M.Wilson in London, where Euclid's text is said to be "antiquated, artificial, unscientific and ill-adapted for a text book" and the theory of parallels "faulty" (Wilson 1868, 361-70) .
This question is dealt with in the same Giornale by Hirst and Hoüel, Brioschi and Cremona (the last with a long letter) and by a short reply by R.Rubini, Wilson's anonymous translator.

Brioschi and Cremona pointed out that Wilson's arguments "are not formidable or essentially new: they are the same ones brought forward in the past centuries by those who were looking for the 'via regia' to learn the elements" and again: the problem of parallels is not solved by having recourse to the concept of direction, as stated by Wilson, but in the sense of non-Euclidean geometry (here the authors mention Beltrami's essay (1868) published in the same journal, where "any obscurity is removed from this argument" [1]). Finally the authors concluded: "in agreement with our learned friend Prof. Hirst according to what he told us during his last visit to Milan, we shall accept Euclid revised, provided it is not Euclid disfigured, and provided it be real geometry and not arithmetic" (Brioschi-Cremona 1869, 52). This line was pursued some time later, when Euclid's text was replaced in the schools by valid books by Italian geometrists (Sanna, D'Ovidio, Faifofer etc.) .

3. Italian unification shared many common elements with the German unification process; mention was made earlier of the close relationships between the two countries in the mathematical field. In Italy however, as opposed to Germany, scientific progress was outstanding virtually on-

1 Beltrami's essay did not succeed in convincing certain mathematicians, such as Genocchi and Bellavitis, who remained staunch and obstinate opponents of the new geometries to their deaths.

ly in mathematics. In the field of physics, for instance, research dealt mainly with mathematical rather than experimental physics, true to a tradition going back to the beginning of the century and strengthened by Beltrami's and Betti's work.

The development of mathematics was certainly favoured by the "schools" which gathered around famous names such as Betti and Cremona; but it was also due to strictly economical influences, because, for a country with limited natural resources and having financial difficulties - such as unified Italy had - the fostering of mathematics, as against other sciences, "offered (...) the advantage of not requiring very expensive equipment" (Candeloro 1978, VI, 295) .

References

Beltrami, E. 1868 Saggio di interpretazione della geometria non euclidea, Giornale di Mathematiche, 6(1868), 284-312

Bottazzini, U. 1980 Tardy's papers and library in Genua, HM 7(1980), 84-85

Brioschi, F. - Cremona, L. 1869 Al signor Direttore del Giornale di Matematiche..., Giornale di Matematiche, 7(1870), 51-54

Candeloro, G. 1978 Storia dell'Italia moderna, Milano.

Casorati, F. 1868 Teorica delle funzioni di variabili complesse, Pavia

Cremona, L. 1873 Elementi di geometria proiettiva, Torino

Loria, G. 1904 Sur l'enseignement des mathématiques en Italie, Verhandl. des dritt. int. Math.-Kongr. (1904), Leipzig, 1905, 594-602

Novi, G. 1863 Trattato di algebra superiore, Firenze

Struik, D.J. 1967 A Concise History of Mathematics, New York

________, 1979 Introduzione all'edizione italiana di Struik, 1967 (to appear)

Volterra, V. 1900 Betti, Brioschi, Casorati, trois analystes..., Compt. Rend. du 2eme Congr. des Math. Paris, 1900, 43-57

________, 1908 Le matematiche in Italia nella seconda metà del secolo XIX, Rend. 4° Congr. Int. dei Matematici, Roma (1908), 55-65

Wilson, J.M. 1868 Euclide come testo di geometria elementare, Giornale di Matematiche, 6(1868),361-70

THE EMPLOYMENT OF MATHEMATICIANS IN INSURANCE COMPANIES IN THE 19TH CENTURY

Horst-Eckart Gross

One important social function of the universities, not only in the Federal Republic of Germany, is the training of highly qualified workers who are actively engaged in practically every part of society. This task has been accomplished to a certain degree by the universities of the 19th century: lawyers, doctors, theologians have already been prepared at the universities for their professional practice at that stage.

The training of highly qualified workers at universities has been remarkably intensified at the 20th century. Towards the end of the 19th century one half per thousand of the population (1) were university students in Germany, and nowadays 20 % of an age group are studying at universities in the Federal Republic of Germany. The prognosis for 1990 is that 11% of the workers of all kinds will hold a university degree (2). This quantity indicates an important characteristic of the proportion of science and society in the 20th century: the direct engagement of science and scientists in industrial enterprises. In the 19th century university graduates have mainly been employed by governmental and ecclesiastical institutions resp. had a so called free-lancing profession like medical doctors and lawyers. In the 20th century ten thousands of scientists are employed in private enterprises (3).

The connection between development of science, training of highly qualified workers and their subsequent employment in industrial enterprises has so far been investigated only in a superficial manner. Individual aspects like difficulties of adaption when changing from university to profession, the connection between "pure" and "applied" research or the activity of scientists in research departments were in the limelight (4). The analysis of the working process of scientists could be of importance in clearing up a certain number of problems, e.g. the relationship between science and production in general, the reaction of science on practical demands, but also questions as an adequate practical interconnection between curricula and the strategy of educational planning.

Especially concerning mathematics, in the Federal Republic of Germany we have the following situation. In 1961 about 2.000 mathematicians out of 9.000 worked in industrial enterprises, in 1970 already 4.000 of altogether 16.000 mathematicians were employed outside schools and universities (5). Judging by the number of students, in the future one third of all mathematicians will be employed in industrial enterprises (6). One can see from this that this quantity - even in the quantitative aspect - can no longer be neglected.

The "scientific community" of mathematicians which is mainly working at universities in the Federal Republic of Germany has three main functions: a) research, b) training of mathematics teachers and c) training of mathematicians for industrial enterprises. The "scientific community" of mathematicians in Germany and later in the Federal Republic of Germany is able to score an enourmous growth which is nevertheless caused by the formal allocation of training functions. In 1864, there had been 42 mathematicians at the universities of the subsequent German Reich and in 1900 their number had already increased to 82. In 1976, the teaching staff at the universities of the Federal Republic of Germany comprised 1.243 mathematicians, they were joined by 1.497

scientific assistants with contracts for a limited time (7). With this development, an extensive personal basis for the development of mathematics as science was given.

In the 19th century, the training function was allocated to the mathematicians at universities. The introduction of mathematics classes at Grammar Schools as well as the legal standardization of school training by the Prussian state were the basis. In the 18th century one was hardly talking about mathematics classes in a real sense. They were developed not before the end of the century. In the Prussian regulation of the year 1735, no mathematical knowledge had been required of graduates of a Secondary School (8). Nevertheless, classes of mathematics developed in schools so that on the basis of the effective development of the profession "teacher of mathematics" the formal introduction of that profession was achieved. This was due to the Prussian edict of July 12th, 1810, by which a special examination for the teaching profession at Secondary Schools given by universities was introduced. By the introduction of this state examination "the subject mathematics was put on the same level with other subjects, ..." (9). The importance of the subject mathematics was emphasized in 1834: those who did not pass the examination were not allowed to enter university, and in the A-level examination mathematics was indispensable (10). The formalization of the training and the intensification of the conditions of access to the universities had considerable effects on the students of mathematics courses at universities. Paulsen wrote the following about mathematics courses and their more and more specialized character: "In the 18th century, universities were open to all students but up to now, they have narrowed their circles more and more, and they have excluded nearly everybody who does not devote himself to special study. The students of the university become nearly exclusively applicants for the _facultas docendi_, whether for academic classes or for the classes at a 'Gymnasium' " (11).

The reasons for this procedure of the Prussian state can hardly be directly and immediatly deduced from the requirements of the productive forces, as the industrial development of Prussia was beginning not before the thirties of the 19th century, and agriculture as well as manufacture and trade had no need of mathematicians at that time (12). The reform of education, however, had been a success of the rising civil powers facing the feudal aristocracy.

Thanks to the reform in education the number of the students of mathematics rose. There are hardly any statistical data, but according to estimations, the number of state examinations in the third decade of the 19th century can be assessed at 20 per year. By the end of the century, the number rose to nearly 300 (13). On the basis of the reform of education, the number of students of mathematics developed without any new regulation.

The situation of mathematicians working in all branches of industry today - certainly not equally distributed but with considerable focal points in the electrotechnical industry and in the insurance (14) - is somewhat different. This situation of today is the result of a historical process in which formal arrangements and the introduction of a special diploma in 1942 have a certain importance, but the actual stimula in this process are much more manifold and complicated and they have hardly been inquired yet. The employment of mathematicians in enterprises began in the insurance in the 19th century. For a long time this remained the only region of employment besides school, approximately up to the end of World War I. During this period, an immense unemployment of teachers had to be recorded, the result of which was that unemployed mathematicians looked for work in other economic branches - and were successful in some cases. Thus a process of diffusion took place although one could not speak of direct demands of mathematical qualifications by the enterprises in most cases. Reports on this process of diffusion were published e.g. by Weiß (15). They show that already in the thirties mathematical qualifications were necessary in different domains. At the beginning of this process of diffusion, however, mathematicians were regarded as academically trained specialists in the first place, and only in the second place their mathematical qualifications were of importance. The development, especially that of the fascist armaments industry required more and more the use of mathematical methods and procedures which the engineers didn't know or were not able to apply, and thus, one arrived at the employment of mathematicians especially in this domain. This is clearly reflected in Kamke's report "To which professions, beyond the school system, mathematicians pass over and what has to be done for them at universities?" (16). The order chosen by him probably corresponds to the distribution of mathematicians to the specific domains in that time. As domains of employment he states: "In the first place, mathematicians in the army, secondly mathematicians in the field of probability and statistics in economy and industry, statistics, finances and insurance, furthermore shortly called economy-mathematicians, in the third place, mathematicians in technique and industry" (17). In 1942, the conditions for the increasing employment of mathematicians in economic enterprises are taken into account by the creation of a special course of study and a special examination. In the sixties of this century, the employment of mathematicians in economic enterprises has reached large quantities due to the electronic data-processing systems and the use of operations-research-methods.

In some essential aspects, the employment of mathematicians in insurance companies in the 19th century antecipates the later employment of mathematicians in other domains. Hence, this is not an isolated and untypical occurrence, and therefore, it seems to be sensible to deal more precisely with this working process. But first some general remarks on the development of the insurance (18):

Already in the Roman Empire there had been predecessors of the

the insurance, to be more exact, there had been predecessors of life insurance. There had also been insurance-like systems in the guilds and corporations af the Middle Age. "As insurance organizations, all of these organizations remained in their starting points. They did not lead to a life insurance in today's sense. Instead of them, other developments were substantial, developments which are connected with the beginnings of the capitalism in the Middle Age." (19). In Germany the insurance, especially the life insurance as the form of insurance with an explicit use of mathematical methods and procedures as well as with the employment of mathematicians started with the development of capitalism in the 19th century. Towards the end of the twenties the first foundations took place and from 1852 to 1857, new life insurance companies were founded within the scope of the general economic impetus. All companies were touched by the economic crisis of 1857 and 1866. They led to centralization and concentration. Another foundation wave - with a strong speculative character - took place in the so-called "foundation years" from 1871 to 1873. Altogether, the growth of the insurance companies from their foundations up to the end of the 19th century is obvious by the development of the number of insurance policies and the insurance capacity represented by them (20):

	Policies	Insurance capacity
1852	46.980	58 Mio Tlr
1860	129.589	138 Mio Tlr
1865	200.627	623 Mio M
1870	348.930	1.008 Mio M
1896	1.181.958	5.122 Mio M
1900	1.475.529	6.404 Mio M

When presenting the development of the insurance companies and especially that of life insurance, the importance of mathematicians in respective enterprises is often being pointed out. But it is seldom dealt with formal qualifications which can hardly be found out for a greater number of mathematicians. In 1894 Kieper stated already that the question, to what degree "managers of insurance companies are pretrained is hard to be answered, and that self-instruction must often have been of great importance" (21). Lorey points out that in the 19th century already, there were qualified mathematicians among the mathematical managers of insurance companies. These mathematicians had incidentally adquired knowledge of the insurance science on their own. To exemplify this he mentions some biographies (22). But at the same time he points out "that you can develop to an eminent actuary even without having completed the studies of mathematics ..." (23).

Another hint is given by Manes by referring to Lorey (though without giving further details): "In Germany there are 59 life insurance companies including the small reciprocal enterprises. If these employed more and more qualified mathematicians instead of other assistants in their account offices, the number of positions will always amount to some hundreds only." (24). And additionally Lexis explains: "Up to the most modern times, insurance companies have always taken their scientifically trained workers from among those wo had originally begun their study with a different aim e.g. who wanted to devote themselves to the mathematical subject or to law." (25).

Without being able to make more detailed and quantitative specifications, one can assume - according to existing specifications - that already in the 19th century mathematicians with a university education had been employed by a considerable number of insurance companies - not accidently caused. This estimation is underlined by the fact that already since 1838, and considerably more intensively since 1895, specific courses concerning actuarial theory have been offered at German universities (26), and that in 1895 the Royal Seminar for Insurance Sciences was established at the university of Göttingen with the active cooperation of Felix Klein. The seminar was established with the aim "to give the opportunity of an adequate scientifical training to those who want to be employed as mathematicians or higher administrative officials in public or private insurance business."(27). Thus, mathematical theory of insurance required a considerable part, there even existed courses exclusively for mathematicians. But it must emphasized that a relatively small number of mathematicians was studying at this seminar; in 1895 only 2 out of 13 students were mathematicians, in 1900, only 13 out of 56 students were mathematicians (28).

The question now is which importance did and does mathematics have in the insurance and which tasks did and do mathematicians have. It should be clear that these are two different kinds of questions: other persons apart from qualified mathematicians are working by using mathematical methods and techniques, and mathematicians do not only have the function of dealing exclusively or even mainly with mathematical problems.

The mathematical basis for the insurance has already been founded in earlier centuries. The first compound interest tables were published by Stevin in 1585. Halley was using systematically the calculation of compound interest with the help of algebraic formula. This work was continued by Jakob Bernoulli among others. The development of the theory of probability was of great importance, and contributions to this have been made by Leibniz, Wallis, Pascal and Fermat. Concepts of great value forthe insurance as "medium life-span" and "average life-span" were specified and defined by Huygens. On this basis the lawyer de Witt in cooperation with the mathematician Hudde wrote the first mathematically well-founded pension calculation. In 1693 Halley published the first mortality table on the basis of which de Moivre put up the calculation of the life annuity. The pension calculation was further developed by Simpson among others (29). Thus, the development of the mathematical theory of insurance was in those days closely connected with the development of mathematics as a whole.

Only in the middle of the 18th century the technique of life-insurance as a separte and special scientific branch, the further growth of which was pushed ahead by the experts of insurance companies, mainly developed in England. In the middle of the 18th century, the "Equitable" was founded, the first life insurance company with a mathematically founded basis. In England, the science of actuarial theory was developed and was given special impulses by the work of Price. We must also stress the works of Euler who wrote 14 papers on probability calculation and its applications. In four papers he especially wrote about mathematical and particularly statistical foundations of life insurance - an indication for the fact that even in the 18th century excellent mathematicians still dealt with the problems of actuarial theory.

In spite of the creation and further development of mathematical foundations, the activities of insurance companies on this basis were not at all self-evident - the insurance of that time often were deceiving enterprises, often connected with bet-communities. An adventurer like Tonti invented the so-called "Tontine", which has for a long time found propagation as a mixture of insurance and bet. Only towards the end of the 19th century, the insurance companies in Germany became "serious" enterprises so that a short-time interest in profit made by deceiving manipulation was not in the foreground of the enterprises' objectives. They wanted long-term business with real efficiencies - this yielded greater profits than deceiveful manipulations. This reasonable view had to be enforced by governmental acts among other things. England was the first country to publish a first order in 1870, the "Life Insurance Companies Act" - after the collapse of two big life insurance companies in the preceeding year. Although governmental provisions were only achieved more than 30 years later in Germany, the enterprises had already begun to put their activities on a solid basis of insurance mathematics. The following items belonged to that basis:

- an annual profit and loss calculation,
- an annual review,
- drawing up of a valuation.

Now as before, the tariff formation was in the focus of attention. "To detect the problems occuring in the tariff construction in an example more easily the simplest form of a life insurance should be examined; that is the one-year death insurance. The sum insured becomes payable if the insuree dies during the following year. When calculating the premium of this risk-insurance, one starts from the hypothesis that for every insuree there is a probability to die in the course of the following year (mortality). That means that among insurees who are equally insured, death is to be expected in a certain percentage, and that there are an equalization of risk when a sufficent number of people are equally insured. Numerically, however, mortality is not known. Estimation is necessary, for which experiences from the past on the frequency of deaths are available, e.g. in the form of mortality tables on which the one-year frequency of mortality with regard to sex and the age that is reached is indicated. Therefore the tariff constructor creates - in this case in the form of probability - a model which initially is theoretical, and which he tries to adapt to reality with the help of empiric numbers. Beside mortality, he also takes interests and administrative expenses into account. This is the basis of calculation. The calculation model is completed with numerical details on interests and costs." (30)

The tariff construction has constantly been improved. Thus, about 60 % of the mathematicians working in the insurance are occupied with tariff constructions today. (31) The proportion between the gross-premium (paid premiums) and the net-premium (disbursed premiums) can be regarded as an indicator for the refinement of mathematical methods of tariff construction: In 1902, insurance companies still kept 64,3 % of the gross-premiums whereas in 1970, the percentage decreased to 33 % (33). There are indications that the proportion was even more profitable for the insurance companies before 1900. Even if the decision on the proportion of gross- and net-premiums is made by the executive board according to the competition and business situation and is not a mathematical question, calculations and forecasts of the mathematicians form an important basis for the decision.

Another important activity of actuaries was that of measuring mortality, somthing which had mainly to be done by the enterprises themselves in the 19th century.The aim was to obtain calculation bases in form of mortality tables but also to discover empiric regularities. In the 19th century this problem had to be solved with only a small statistical sample.

In England, the registration of all cases of death were decreed only in 1836, and in Germany, the Grand Duke of Oldenburg accomplished a pioneer's work in 1867: for the first time the dead were classified with regard to their year of birth, age group, year of death. With the improvement of the official statistics more predicative mortality tables and with that, better bases for the insurance activities could be established.

In the sixties and seventies of the 19th century, the calculation of premiums for the arranging agent brought about great polemics among the actuaries. These premiums were continously rising until they amounted to 2,5 % of the sum insured in the seventies. Not few mathematicians took part in this dispute.

A number of mathematicians were also employed because of direct competition reasons: To establish tendency statistics and the more extensively in offices of competition. The first tendency statistic appeared in the year 1852, then it became more and more usual to use mathematics in order to lower competition. Braun judges these statistics as follows: "These reports did not content themselves with a mere schedule of numerical data published by the individual insurance enterprises, but they formed certain proportionality factors from that which gave the layman the impression that entirely particular advantages arise for the one part of the companies and disadvantages which weigh heavily, would manifest themselves for the other part of them." (34) These practices soon belonged to ordinary business activities. Ehrenzweig said about that in 1895: "The mutual calumny has already gone so far that diverse societies have established 'competition offices'. One of these made itself most unpleasently felt by the thorough study of competing societies and the search for their real or alleged weak points which was done by especially for this appointed mathematicians. From time to time the result of these inquiries is imparted to their agents to make frank use of it. Tendency statistics is here in bloom and for the ethical insurance this is a disgraceful spectacle performed intra et extra muros, indeed." (35) Today, insurance enterprises still mantain departments where competition is examined.

Form the first beginnings of the insurance it was another important task to prepare the balance sheets and to prepare the distribution of the profits. This refers to the following questions: "The model for tariff calculation will not correspond to the actual course of things, among others because it was chosen too conservatively. The adaption of premiums which have first been made on the basis of a model results from profit-sharing within the scope of which the surplus made in consequence of the actual course is for the benefit of the insurees. In the broadest sense, profit-sharing is the repayment of premiums which have unnecessarily been paid. The share in capital amounts, which insurance companies havc made beyond interest by profitable capital investments, also belongs to this. By the way, for the time being, this is the most important surplus source. The mathematicians' task is to lay down the profit system according to which this surplus are distributed among individual insurance companies and how this surplus is used. There is a whole scale for the distribution systems which reaches from the simplest form, the so-called mechanical dividend in a percentage of the premium, up to the most complicated forms where the surplus is distributed according to its emergence from different surplus sources as fair as possible. There are numerous possibilities for the use of the share in profits (balancing of premiums, collection of bearing interest, improvement of benefits). The system and the way it is used (e.g. according to the amount of the sum insured) are to be chosen in a fair, simple and marketable manner." (36)

Numerous other tasks could be mentioned where mathematical knowledge was used. Requirements of actuaries in the mid 19th century are shown in the examination demands for different kinds of members at the Institute of Actuaries of Great Britain. In spite of the existing differences between an "actuary" in England and an insurance mathematician in Germany one can generally assume similar fields of practice and with this similar qualification demands, at least in the field of mathematics. Thus, the following qualification profile could also have been valid for mathematicians in German insurance enterprises.

Students had to prove the following knowledge:

1. Arithmetic and algebra, theory and the use of logarithms, elements of probability theory.
2. Elements of difference calculus including interpolation and summation, elements of differential and integral calculus, excluding trigonometrical problems.
3. Compound interest and pension calculus, construction and use of corresponding tables.

Passive members had to prove the following knowledge:

1. Life annuity and insurance, construction and use of the corresponding tables.
2. Policy grouping for premium reserve calculation, premium reserve index.
3. Accounting, especially considerations of life insurance and further knowledge.

Full members had to possess knowledge in:

1. The method of mortality, illness, accident etc. statistics, setting up and balance of these tables.
2. The valuation of the obligations of life insurance and employer liability companies.
3. Principles and methods of surplus distribution.
4. Gross-premiums of life insurance, pensions, and so on.
5. Special premiums for inferior lives, dangerous professions and stay in the tropics and adequate material for their definition. (37)

When extensive theoretical knowledge was demanded, it has to be noted that compared to the development of mathematics at universities, mathematics used in practice often had elementary character. In 1903 this has also been remarked by von Bortkiewicz in his essay on the classes for actuarial theory at German universities at the IV. International Congress of Insurance Science in New York:

"The tasks an actuary is confronted with in practice, as e.g. the setting up of balances and premium tariffs can be solved on the basis of certain calculations which are mainly of elementary nature. They imply the application of formulas which essentially belong to lower algebra. As you know, practice avoids maling use of the higher parts of life insurance calculation, as especially of the risk theory, and everything else connected to practice does not even demand the knowledge of elementary probability calculus which can doubtlessly be left out when developing premiums and reserve formulas. An absolutely secondary practical importance is due to the development of series which can be found in life insurance calculation textbooks and which lead to certain approximation formulas. If you also bear in mind that the basic thought, from which every calculation in the domain of life insurance starts, is very simple, you might think that everybody who has got some experience in calculation - not to speak of professional mathematicians - will be able, if he only once comprehends this basic thought, to discover the right way to the problems' solutions which emerge in the insurance business, and all this without special instructions but only on the strength of his mathematical knowledge. But in reality this is not the case. First of all, as is taught by the history of life insurance calculation, with regard to the particular question you do not always easily find the right, not to speak of the shortest derivation of the relevant formula. Don't you remember how one had to struggle with the calculation of life annuity's value on lives related with each other until the only correct method was found? There was a lack of excellent scientific authorities or they took a circuitous route where we choose a correct and shorter method because the necessary patterns which we have easily taken over are at our disposal. Then personal experiences which can be gained as a lecturer or otherwise with trained mathematicians, show that they pretend an air of awkwardness in the beginning when they are wanted to set up a premium or reserve formula for an insurance combination which they have not been taught up to then." (38) The advantage of theory developed at universities becomes clear compared with the methods used in practice. This aspect of the relation between theory and practice is generally known. However, less known is the fact that this margin of theory does not turn the use of fundamental theoretical knowledge into simple and uncomplicated proceedings. Even the use of fundamental mathematical proceedings and methods call for an effort, thought, and needs a working process which can doubtlessly and just because of this be called a scientific working process. The mode and method of work prove a working process to be a scientific or a non-scientific one, but not the distance of employed and developed theory.

Another important aspect of the mathematicians' working process in the insurance is its interdisciplinary character. Concerning this, von Bortkiewicz says: "You have to take into account that, although the actuarial theory is something which exists for itself beside the insurance economy and insurance right, it often comes close to these two fields of knowledge, and it could easily be shown that certain legal constructions being somewhat abortive and inadequate as well as certain insensitive demands, with which the insurance is confronted by the national economical side, originate from a lack of familiarity with the theorems of life insurance calculation." (39)

In practice, the concrete problem is in the foreground during the treatment of which mathematical methods are only a sort of aid,of tool, even for the mathematician. That is why there is made only a relatively small use of mathematics - but it can only be called small when it is compared with the dealing with mathematics which is to be found at universities. When applying mathematics in practice, even the occupation with non-mathematical questions represents an important aspect of precisely this application of mathematics.

Referring to the study of the working process of actuaries in the 19th century, it can be concluded that for an analysis of the mode of action and the function of mathematics, the study of the working process as a part of social history of mathematics will be of interest. An integral study of the working process in this sense seems to be more promising than the examination of individual aspects like difficulties of adaption or the belonging to a formally defined "scientific community".

Notes and references:

1 Paulsen,F. Die deutschen Universitäten Berlin 1902,p.165
2 Teichler,U. Higher Education and Employment in the Federal Republic of Germany Kassel 1978,p.5 (Working paper of the scientific center of Profession- and University Research)
3 This has also long-range consequences for the social situation of university-graduates. But it is not possible to deal with this aspect in this paper.
4 See e.g. the survey article of Mulkay,M.J. Sociology of the Scientific Research Community, in: Spiegel-Rösing/de Solla Price (eds) Science,Technology and Society London/ Beverly Hill 1977 p.93
5 Specifications according the census and profession census of the Statistical Federal Bureau of the years 1961 and 1970.
6 Out of 32.358 mathematics students who immatriculated in the summer 1979, 42 % studied to get a non-teaching degree
7 See Gross,H.-E.,Das sich wandelnde Verhältnis von Mathematik und Produktion, in: Plath/Sandkühler (eds) Theorie und Labor Köln 1978 p.253
8 Timerding,H.E. Die Verbreitung mathematischen Wissens und mathematischer Auffassung (III.Teil,1.Abteilung von "Die Kultur der Gegenwart") Leipzig/Berlin 1914 p.110
9 Timerding loc.cit. p.114
10 Timerding loc.cit. p.119
11 Paulsen loc.cit. p.531
12 Cf. Mottek Wirtschaftsgeschichte Deutschlands Berlin(GDR) 1969,p.76 and p.130
13 See Project "Mathematik in der Industriegesellschaft", Berufs- und Ausbildungsstatistik im Fach Mathematik, Mimeo Bielefeld 1975 p.29
14 A survey about the actual working areas is given by the empirical studies of the Project "Mathematik in der Industriegesellschaft" at the University of Bielefeld, published in the Materialien zur Analyse der Berufspraxis des Mathematikers vol.1(1971) to vol.25(1980)
15 Weiß,E.A. "Mathematiker im Volksleben" Deutsche Mathematik 2(1937)p.379
16 Kamke,E. "In welche Berufe gehen Mathematiker außer den Schuldienst noch über, und was muß auf den Hochschulen für sie geschehen?" Jahresberichte der DMV (Deutsche Mathematiker Vereinigung) 47(1937)p.250
17 Kamke,E. loc.cit. p.250
18 When representing the working process, it is mainly referred to Braun,H. Geschichte der Lebensversicherung und der Lebensversicherungstechnik Berlin 1925, second unchanged edition Berlin(West)1963
19 Braun,H. loc.cit. p.16
20 Specifications according to Braun,H. loc.cit. p.168,p.269 and p.350
21 Specifications according to Lorey "Das Studium der Versicherungsmathematik" Zeitschrift für die gesamte Versicherungs-Wissenschaft 22(1922) p.283

22 Lorey loc.cit. p.283
23 Lorey loc.cit. p.283
24 According to Manes,A. *Versicherungswissenschaft an deutschen Hochschulen* Berlin 1903 p.30
25 Manes quoted loc.cit. p.30
26 See Manes loc.cit. p.8
27 See Manes loc.cit. p.30
28 See Manes loc.cit. p.16
29 More detailed specifications see Braun,H. loc.cit.
30 Ellger,W. "Der Mathematiker in der Lebensversicherung" *Materialien zur Analyse der Berufspraxis des Mathematikers* vol.6 Bielefeld 1972 p.64
31 Project "Mathematik in der Industriegesellschaft" *Mathematiker mit abgeschlossenem Hochschulstudium in privaten Versicherungsgesellschaften in der BRD 1971-1973* Mimeo Bielefeld 1974, especially p.8
32 According to Pritzkoleit *Wem gehört Deutschland* Wien/München/Basel 1957 p.247
33 Handelsblatt December 16th, 1971
34 Braun,H. loc.cit. p.66
35 *Ehrenzweigs Assekuranz-Jahrbuch* (Wien) 16(1895)
36 Ellger,W. loc.cit. p.66
37 According to Braun,H. loc.cit. p.234
38 The text of the lecture can be found as an annex in Manes loc.cit. p.64
39 von Bortkiewicz, p.66 in Manes,loc.cit.

PART III

INDIVIDUAL ACHIEVEMENTS IN SOCIAL CONTEXT

INTRODUCTION

Henk Bos

The three articles in this section raise the issue of social influences upon the concepts and methods of mathematics itself. In doing so they point to new modes of understanding and explaining developments of the internal structure of mathematics - developments which until now have been understood almost exclusively in terms of the internal logic of mathematical development.

All three articles concentrate on persons and their achievements in mathematics. Thomas Hawkins writes about Frobenius and Killing and their contribution to matrix theory and to the early theory of Lie-algebras respectively. Albert Lewis studies Grassmann and his programme for a new science of extension presented in the *Ausdehnungslehre* of 1844. David Bloor discusses Hamilton's contributions to mathematics and his views on the nature of mathematical concepts and symbols, contrasting these views with those current among Cambridge mathematicians at the time.

In Hawkins' article the central theme is the influence of a mathematical *school* on the achievements of those who have been trained or who work within it. The school in question is the Berlin school. Both,the attitudes it stood for and its views on what is important in mathematics were set by Weierstrass. Frobenius and Killing worked outside Berlin, in fields not central to the interests of the Berlin school. Both departed from the directions of research current in their fields; those new approaches can be understood as resulting from the influence of the Berlin school's style, which Frobenius and Killing took over in their student days at Berlin.

Lewis discusses an influence from outside mathematics, namely the influence of Schleiermacher's philosophy on Grassmann's Ausdehnungslehre. There are similarities between Grassmann's new programme for a new science of extension as a base for all mathematics and Schleiermacher's philosophical and theological views. In particular there is a common interest in dialectics, polarities,and the role of the individual. These similarities suggest an influence which can be traced to Grassmann's earlier contacts with Schleiermacher's ideas. Awareness of this influence helps us to understand Grassmann's work and its reception in Germany.

In his article on Hamilton, Bloor traces the philosophical and political ideas which influenced Hamilton's views on the nature of mathematical concepts and symbols. These ideas are apparent in Hamilton's studies on the concept of number, in his creation of quaternions, and in his opposition to views about the nature of mathematical symbols current among Cambridge mathematicians. But Bloor adds a further level of understanding with his discussion of the social context of Hamilton's mathematics. Not only were Hamilton's concepts in mathematics influenced by the social context; these concepts also served a function transcending the strictly mathematical realm. They carried a social message reflecting Hamilton's views on how society should be structured, in the same way as ideas about nature often reflect and enhance views on the arrangements in society.

The material and the ideas in this section, especially the suggestions for new types of explanation, led to intense and fruitful discussions during the workshop. No doubt the articles will stimulate a continuation of that discussion.

HAMILTON AND PEACOCK ON THE ESSENCE OF ALGEBRA

David Bloor

In the *London Review* of 1829 the mathematician Baden Powell observed

> that the most violent controversies have arisen out of the speculations of mathematicians; and that even at the present day, and among the greatest mathematical luminaries of the age, considerable difference of ideas prevails as to the relative value, importance, and even validity and correctness of different methods of investigation. (1)

Looking around him Baden Powell saw evidence for a variety of different schools of thought and, he noted, 'a new school seems to be gaining ground'. (2) He was referring to the Cambridge advocates of 'symbolical algebra' with their new continental methods of analysis. Led by Peacock, Babbage, Whewell and Herschel this group differed in many of their judgements of value and validity from other mathematical luminaries. For instance, they had a protracted difference of opinion with the celebrated Irishman Sir William Rowan Hamilton about the essential nature of algebra. The Cambridge group defined algebra as 'the science of general reasoning by symbolical lan-

1. Baden Powell, *London Review*, vol.1, no.2, 1830, p.467.
2. *Ibid*., p.467.

guage'. (3) They might be called 'formalists', whilst Hamilton, as we shall see, could be called an 'intuitionist'. (4)

This divergence of opinion about the essential nature of algebra will be my topic. I am interested in why men who were leaders in their field, and who agreed about so much at the level of technical detail, nevertheless failed to agree for many years about the fundamental nature of their science. I shall propose and defend a sociological theory about Hamilton's metaphysics and the divergence of opinion about symbolical algebra to which he was a party.

1. Apart from his great achievements in optics and mechanics Hamilton is remembered for two contributions to algebra: he had the idea of representing complex numbers (a + ib) in the form of ordered pairs of real numbers (a,b); and he is famous for his discovery of quaternions. These are hyper-complex numbers of the form (a + ib + jc + kd) where, loosely speaking, the i, j and k are 'imaginaries'. (5)

3. George Peacock, A treatise on algebra, 1830, p.1.
4. Using these labels may be anachronistic, but it is not without point or precedent. For example, M.R. Cohen called De Morgan a formalist - M.R. Cohen, Reason and Nature. An essay on the meaning of the scientific method, Free Press, Collier-Macmillan, London, 1964, p.184; and the Dutch intuitionist Brouwer says that mathematics is 'inner architecture' which is grounded in consciousness whose 'initial phenomena is a move in time', see L.E.J. Brouwer, 'Consciousness, philosophy and mathematics', Proc. of Xth Ann. Cong. in Phil., Amsterdam, p.1235 and p.1249.
5. For a general assessment of Hamilton's place in the history of mathematics see, for example: F. Cajori, A history of mathematics, New York, Macmillan, 1894, esp. pp.318-319; and M.J. Crowe, A history of vector analysis: the evolution of the idea of a vectorial system, Univ. of Notre Dame Press, Notre Dame, 1967.

Hamilton's metaphysical interests placed him securely in the Idealist tradition. He adopted the Kantian view that mathematics is synthetic a priori knowledge. Mathematics derives from those features of the mind which are innate and which determine a priori the general form that our experience must take. Thus geometry unfolds for us the pure form of our intuition of space. Hamilton said that if geometry was the science of pure space, then algebra was the science of pure time. In 1835 he wrote:

> My metaphysical meditations upon Algebra have been for some years settling into a conviction that Algebra is the science of Pure Time. (6)

And he added that

> Among professed Algebraists, few have failed, indeed, to introduce some passing illustrations from the thought of Time; and Newton's theory of Fluxions was mainly founded on that thought: but .... among those who reason at all upon the subject, opinions seem to be of late converging on this point, that Algebra is merely a language. (7)

Hamilton was convinced that he was able to make his innovations in algebra precisely because he conceived it as the science of time. Thus he said:

> The quaternion [was] born, as a curious offspring of a quaternion of parents, say of geometry, algebra, metaphysics and poetry .... (8)

These, he said, 'led me to strike out some new lines of research, which former methods had failed to suggest'. (9)

6. Robert Graves, Life of Sir William Rowan Hamilton, London Longmans, 1832, 3 vols. e.g. vol.II, p.146.
7. Ibid., vol.II, p.147.
8. Thomas Hankins, 'Triplets and triads: Sir William Rowan Hamilton on the metaphysics of mathematics', Isis, vol.68, no.242, June 1977, pp.175-193, p.176 (from an unpublished letter of 1855).
9. Ibid.

Historians have not been so sure. Bell dismissed the metaphysics as irrelevant to the mathematics, (10) and Whittacker exhibits quaternions as arising out of the problem of how to generalise the two-dimensional geometrical representation of complex numbers to three dimensions. (11) Recently there has been a shift of emphasis. Crowe suggested that metaphysics played a decisive role as a support for the discovery of quaternions once it had been made, and Hankins has argued strongly that Hamilton's metaphysical theories 'helped direct his mathematical researches'. (12)

10. E.T. Bell, Men of mathematics, New York, Simon and Schuster, 1965, p.358 (quoted in Hankins, 1977, p.191).
11. E.T. Whittacker, 'The sequence of ideas in the discovery of quaternions', Proc. Roy. Irish Acad., vol.50, 1944-5, pp.93-98. The geometrical representation of complex numbers usually known as the Argand diagramme was introduced into Britain by Warren. J. Warren, A treatise on the geometrical representation of the square root of negative quantities, Cambridge, 1828.
Hamilton was aware of this book and registers his debt, see W.R. Hamilton, Philosophical magazine, vol.XXV, 1844, pp.489-95.
12. Crowe, op.cit. p.95; Hankins, 1977, p.176; and also Thomas Hankins, 'Algebra as pure time: William Rowan Hamilton and the foundations of algebra', in P.K. Machamer and R.G. Turnbull (eds), Motion and time, space and matter, Ohio State Univ. Press, 1976, pp.327-59, p.328, p.332, and p.335. I am heavily indebted to these valuable papers and some, though by no means all, of the material I quote was first encountered in Hankins' discussion. The main exception concerns Hamilton's politics, a subject which falls totally outside the scope of Hankins' papers. The justification for going over some of the same ground is that I want to show how the pieces of the jig-saw can be fitted together to form a different picture.

I shall not try to contribute to this difficult question but shall confine myself wholly to the origin and meaning of the metaphysical doctrines about algebra. Should it transpire that this metaphysics is indeed relevant to the technical mathematics, then my ideas may help to illuminate these matters as well. (13)

2. Hankins has traced for us the detailed chronology of Hamilton's involvement with Idealism. It began with his reading of Madame de Stael's account of German thought, and his taking careful (shorthand) notes from Thomas Carlyle's anonymous article in the Foreign Review of 1829. It grew with his meeting with Coleridge in 1832 and 1833. Coleridge was himself an enthusiastic proponent of German Idealism and both his poetry and philosophy greatly impressed Hamilton. We then hear of Hamilton's struggle to acquire and translate the Critique of Pure Reason. Finally we are shown the close similarity between Coleridge and Kant's doctrines and Hamilton's own beliefs as revealed in his letters and publications. Hankins establishes that 'The period of his greatest interest in their philosophies coincided with his work on the foundations of algebra'. (14)

13. There is some reason to think that general commitments of the kind I shall consider can have very technical ramifications. The twentieth-century debate between formalists and intuitionists is a case in point. Even in the present instance there are some indications of this kind. Thus, the formalists' rejection of the doctrine of limits was a technical stance within mathematics proper, and certainly seems to have been connected with their desire for algebraic and symbolical 'purity'. Hamilton was more sympathetic to older fluxional ideas (which used limiting processes) as well as to geometry generally.
14. Hankins, 1976, p.332.

In his second paper Hankins enlarges on the similarities between Hamilton and Kant by discussing the _Critique of Judgement_. We also hear more of Coleridge. Instead of being presented as a mere channel through which Hamilton received the (somewhat distorted) influence of Kant, the poet himself now comes to the fore. Hamilton's Idealism, we are told, 'had its origin largely in the philosophy of Coleridge'. (15) It was from Coleridge that Hamilton derived his general picture of mind and nature as a set of oppositions which develop through a process of synthesis. In _The Friend_ Coleridge had announced his 'universal Law of Polarity' saying that 'Every Power in Nature and in Spirit must evolve an opposite, as the sole means and condition of its manifestation; and all opposition is a tendency to reunion'. Hamilton had written notes on this, casting it into the language of power and resistance, concluding that 'Existence is manifested by the struggle between two opposite tendencies', identifying these as 'the tendency to change and the tendency to continuance'. (16)

There is, however, a political and social or practical dimension to this Idealist speculation which has not, perhaps, received the attention it deserves. Let us look in more detail at one of the acknowledged sources of Hamilton's metaphysics. Consider again the important article that Carlyle wrote in 1829. (17) It reviews the work of the strange, mystical poet Friedrich von Hardenberg, known as Novalis. It would be a mistake, says Carlyle, to treat Novalis as a dreamer whose works reveal nothing but a peculiar state of mind

15. Hankins, 1977, p.186.
16. Graves, vol.1, p.439.
17. Thomas Carlyle, Art.V, _The foreign review and continental miscellany_, vol.IV, 1829, pp.97-141.

induced - as some commentators would have it - by the death of Sophie, his beautiful child-sweetheart. Rather than being purely personal, Novalis's attempt to Romanticise the world is part of a broader change in attitude. It is nothing less than an expression of German Idealism and *this*, insists Carlyle, is not an other worldly doctrine at all. In fact, it is intensely practical in its significance:

> The reader would err widely who supposed that this Transcendental system of Methaphysics was a mere intellectual card-castle, or logical hocus-pocus, contrived from sheer idleness, and for sheer idleness, being without any bearing on the practical interests of men. On the contrary ...... it is the most serious in its purport of all Philosophies. (18)

Carlyle then goes on to explain precisely how Idealism has a practical bearing. First of all it removes a stumbling block to theology and banishes the black spectre of Atheism. If Idealism is true, 'the old hostility of Matter is at an end, for Matter itself is annihilated'. By making matter dependent on mind, rather than something in its own right, Idealism removes the threat of a rival conception of Reality. There can be no ultimate explanations other than those offered by religion. It pulls the ontological rug from beneath the feet of those who would set themselves in opposition to spiritual authority.

Second, and more specifically, the knowledge of nature provided by the sciences is given a subordinate status. According to Idealism the laws of our own mind are constitutive of nature, and this means that

> all inductive conclusions, all conclusions of the Understanding, have only a relative truth, are true only for *us*, and *if* some other thing be true. (19)

18. *Ibid*., p. 116.
19. *Ibid*., p. 117.

The relativity and dependence of our scientific knowledge is then explained more fully when Carlyle reveals that the Understanding is but one of our mental faculties. There is a *higher* faculty which transcends the Understanding and gives us contact with non-relative and non-dependent Absolutes.

> We allude to the recognition, by these Transcendentalists of a higher faculty in man than Understanding; of Reason (*Vernunft*), the pure, ultimate light of our nature, wherein as they assert, lies the foundation of all Poetry, Virtue, Religion ; things which are properly beyond the province of the Understanding. (20)

Here, says Carlyle, we reach the true object of Novalis's work: to preach and establish the 'Magesty of Reason'. With a burst of social metaphor Carlyle endorses Novalis's desire that Reason should

> conquer .... all provinces of human thought, and everywhere reduce its vassal, Understanding, into fealty, the right and only useful relation for it. (21)

Carlyle was right. This theory of the relation of Reason to Understanding really was a *practical* doctrine. It was practical in the sense that it had a use, and that use was social and political: it permitted the knowledge of nature to be ranked below the knowledge of Poetry, Morality and Religion. By mapping science and morality onto different faculties of the mind, and then putting these faculties in an hierarchical relationship, Carlyle could build that ranking into the nature of things.

This technique of using the Idealist account of mind and reality is equally evident in Coleridge and will help us to grasp Hamilton's employment of the same tactic. Coleridge

20. *Ibid*.
21. *Ibid*., p. 118.

was a Tory propagandist. (22) His Idealism was a vehicle and justification for his theory of politics and society. Like the other Lake poets, Wordsworth and Southey - all of whom Hamilton knew personally - Coleridge turned against the French Revolution which had raised the enthusiasm of his youth. (23) His earlier political radicalism was replaced by a reaction against the rationalistic, materialistic and individualistic ideologies associated with the events on the Continent. Jacobinism, thought Coleridge, was the consequence of following 'the universals of abstract reason'. (24) It was the human understanding 'usurping the name of Reason'. (25) I assert, says Coleridge, that

> the understanding or experiential faculty, unirradiated by reason and the spirit, has no appropriate object but the material world in relation to our worldly interests. (26)

It was this exclusive attention to wordly interests that had been responsible for the revolutionary calamities of the past, as well as for the social and industrial discontents of the present. These, says Coleridge, are

22. C. Brinton, The political ideas of the English romanticists, Oxford University Press, 1926; A. Cobbam, Edmund Burke and the revolt against the eighteenth centry: a study of the political thinking of Burke, Wordsworth, Coleridge and Southey, London, Allan and Unwin, 1929, esp. Ch. VI; C.R. Sanders, Coleridge and the broad church movement, Durham, North Carolina, Duke University Press, 1942; R.W. Harris, Romanticism and the social order, 1780-1830, London, Blandford, 1969; B. Knights, The idea of the clerisy in the nineteenth century, Cambridge, University Press, 1978, esp. Ch.II.
23. On Wordsworth see G. Dodd, 'Wordsworth and Hamilton', Nature, vol.208, 1970, pp. 1261-63.
    On Southey and Hamilton see, for example Graves, vol.I, p.270, p.283, p.390.
24. S.T. Coleridge, Statesman's manual, quoted Cobbam, p.167.
25. Coleridge, Statesman's manual, quoted Knights, p.49.
26. Ibid., p.49.

> resolvable into the imbalance of the commercial spirit in consequence of the absence or weakness of counter weights. (27)

The commercial spirit was eroding social relationships. Old bonds of duty and reciprocal obligation were being dissolved into an atomised individualism.

So rationalism leads to materialism, and materialism leads to individualism, and this leads to revolution. (28) No wonder that Coleridge was able to tell Hamilton that it was pantheism not popery that was the threat of the age. Pantheism makes God the soul of the world and denies his transcendence. As Coleridge put it, if G = God, and W = the World, then pantheism says: G - W = 0. (A piece of mathematics that Hamilton declared to be 'perfect'.) (29) By implying that God has no existence over and above His creation it makes nature and matter itself divine - and _that_ is just materialism again, the philosophical justification for a Godless and unstable society. (30)

27. Coleridge, _2nd Lay Sermon_, quoted Knights, p.66.
28. From Coleridge's table talk, quoted by Harris, p.227.
29. Graves, vol.III, p.238.
30. Pantheism has a long association with political radicalism. For instance, the belief that God is the soul of the world was adopted, and imputed to, the radical sectaries of the civil war period. It was a target of attack by Robert Boyle and Isaac Newton, and seems to have been one of the main reasons why the corpuscular philosophers adopted the thesis that matter is passive and cannot move itself. See, for example J.R. Jacob, 'Boyle's atomism and the Restoration assault on pagan naturalism', _Social Studies of Science_, vol.8, 1978, pp.211-33; M.C. Jacob, _The Newtonians and the English Revolution, 1689-1720_, Ithaca, Cornel Univ. Press, 1976. For a valuable account of this work which brings out its theoretical significance for sociological theories of knowledge see S. Shapin, 'The social uses of science, 1660-1800', _in_ R.S. Porter and G.S. Rousseau (eds) _The ferment of knowledge : changing perspectives in eighteenth century science_, Cambridge, C.U.P. (forthcoming).

Coleridge's long preoccupation with these problems culminated in his 'On the Constitution of the Church and State according to the Idea of Each', which sums up ideas already broached in The Friend and the Lay Sermons. (31) Here he explains how the insights of Idealism must be embodied in our social arrangements. Society should rest on the truth that the 'organic and living ..... whole is prior to the parts', and then human nature would achieve genuine unity 'through the godlike form of the state'. (32)

Using the analogy of the two poles of a magnet Coleridge identified two opposed but interdependent 'interests' in society. These were the landed interests representing the forces of permanence, and the commercial interests representing the forces of change. Both were necessary for social health, but they had to be in a properly balanced relationship to one another. To achieve harmony some principle of synthesis was required. This would have to be the institutional embodiment of those higher values which alone could sustain social unity. Proceeding dialectically Coleridge argued that the common ground of the two social elements so far described was that they were property interests. The third element in the constitution, which would synthesize them, would have to be a non-property interest. The 'Propriatage' would have to be balanced by the 'Nationality'. The new unifying group was to be called the 'Clerisy'.

> I hold it a disgrace and calamity to a professional statesman not to know and acknowledge, that a permanent nationalised, learned order, a national clerisy or Church

31. S.T. Coleridge, 'On the constitution of the church and state according to the idea of each', 1830.
32. Quoted by Knights, p.41 from a letter of Coleridge's to Lord Liverpool.

> is an essential element of a rightly constituted nation, without which it wants the best security alike for its permanence and its progression. (33)

Coleridge, like Carlyle, was relating mind and knowledge and society and nature. Clearly his model of society is identical to his universal Law of Polarity, running through all Spirit and Nature, while the Faculty of Reason is the channel through which moral truths are to be transmitted to the men of knowledge - the Understanding merely catering for worldly interests. The usurpation of Reason would be at an end when it could be represented and defended by the Clerisy. The social role of the man of knowledge was, therefore, to mediate and control the lower-level interests, irradiating them, in Coleridge's words, with spirit.

3. These same social themes were taken up by Coleridge's followers and those who were, like Hamilton, members of his circle. Thus in 1832 Hamilton's friend, Aubrey De Vere, a fellow Coleridgean, could write to the mathematician of the 'selfishness, the vanity, the drivelling infidelity, the materialism, that has been corrupting the principles and habits of the people'.Prompted by an encounter with some advocates of utilitarianism, or what he ironically calls 'the enlightened principles of modern philosophy', he denounces contemporary society as 'mean and selfish to an inconceivable degree' and wonders how the poets and true philosophers of the day can 'meet the spirit of Democracy and Innovation'.'I confess', he says,'the more I think about politics the more desponding I become'. (34)

33. Quoted by Knights, p.37, from 'On the constitution......'
34. Graves, vol.I, p.618.

This was but an amplified version of the very message that Hamilton had earlier conveyed to De Vere when commenting on his fellow scientists and the state of science. In March and April of 1832 Hamilton had visited London and Cambridge, meeting Peacock, Babbage, Herschel, Whewell, Airy, Ivory, Sedgwick and others. He also fitted in a visit to the Highgate sage as well, but chose to keep quiet about this when in scientific company and 'abstained wonderfully from talking of Coleridge' (35) On his return he wrote to De Vere admitting that the vigorous company of his new Cambridge friends had raised his spirits somewhat despite 'my opinion respecting their habits of thought or thoughtlessness on the subjects which interest me most'. He referred to the fact that the English scientists were 'winning to themselves mansions above the earth, though beneath the highest heaven', observing with regard to metaphysics that, alas, 'the champions of science are not her champions also'. The one exception was Whewell: 'I thought with delight that I perceived a philosophical spirit more deep and true than I had dared to hope for'. (36)

The significance of these remarks becomes clear in the light of Hamilton's reaction to the visit of the astronomer G.B. Airy in 1831. The encounter was not quite as bad as Hamilton had feared. 'But on the whole', Hamilton wrote to Viscount Adair

> his mind appeared to me an instance, painful to contemplate, of the usurpation of the understanding over the reason, too general in modern English Science. (37)

35. Graves, vol.I, p.551.
36. Graves, vol.I, p.553.
37. Graves, vol.I, p.444.

Airy had declared to Hamilton 'playfully perhaps, but, I think, sincerely' that the highest achievement of man was the Liverpool to Manchester railway. As a symbol of the painful imbalance between Understanding and Reason this is revealing: for it concerns an excess within science of precisely that commercial spirit to which Coleridgeans declared themselves opposed. This was what 'chilled' Hamilton, who, in the language of Carlyle and Coleridge, cried out to his friend Adair:

> When shall we see an incarnation of metaphysical in physical science! When shall the imagination descend, to fill with its glory the shrine prepared for it in the Universe, and the understanding minister there in lowly subjection to Reason! (38)

How was this incarnation to be brought about? One way was by the exclusion of materialistic doctrines from science. In 1832 recalling Coleridge's strictures against atomism Hamilton declared such theories to have merely 'subjective value'. Invoking the subordinate status of the Understanding he toyed with the idea of accepting that atomistic theories were

> a fit medium between our understanding and certain phenomena: although objectively, and in the truth of things, the powers attributed to atoms belong not to them but to God. (39)

Two years later he described in a letter his pleasure in meeting Faraday and hearing how 'the most distinguished practical chemist in England has been led to almost as antimaterialist view as myself'. Faraday, says Hamilton, found 'the conception of matter an encumbrance and complication in the explanation of phenomena'. The idea that was to replace 'those little bulks or bricks, of which so many fancy the outward

38. Graves, vol.I, p.444.
39. Graves, vol.I, p.593.

world to be built' by that of immaterial 'power'. (40)

There is an ascending scale here. What appears to be matter is really power, and power is an attribute of God: it is in fact His Will. This is proved to Hamilton by the phenomenon of miracles. At first sight the predictable regularity of nature might make it appear to have no likeness with the exercise of willpower, but

> The experience of miracles makes visible the before unseen analogy of this power to will, by giving examples of the interruption in the usual connexions of phenomena or sequences of sensations. (41)

But he went on:

> Miracles do more; they show that the Being or power which the study of our sensations has led us to acknowledge as the physical governor of the universe is also the moral governor, the power which produces in us involuntary emotions of remorse or peace. (42)

If we put Hamilton's views of mathematics in <u>this</u> context we can see at once what he was doing by relating algebra to our intuitions of pure time. The essence of algebra was given a direct association with the Reason, with what was prior to and determined the form of experience. At the same time it was thereby put in close proximity to our insights into moral truths and their divine origin. In a word, Hamilton was irradiating algebra with spirit.

Now the sacred must always be kept apart from the profane. Hamilton's definition of a True Science allowed him to put an appropriate distance between algebra and more mundane or material forms of understanding. A science of algebra, properly so-called, says Hamilton, must be

> strict, pure and independent; deduced by valid reasonings from its own intuitive principles; and thus not less an

40. Graves, vol. II, p.96.
41. Graves, vol.I, p.413, (from a memorandum of 1831).
42. Graves, vol.I, p.413.

> object of [a] priori contemplation than Geometry, nor less distinct, in its essence, from the Rules which it may teach or use, and from the Signs by which it may express its meaning. (43)

In its essence algebra is neither concerned with useful rules nor with written signs and symbols. Rather, its essence was derived from the laws and constitution of the mind itself - and the most exalted part of the mind at that.

For the discipline of mathematics this 'spiritualisation' had the consequence that algebra dealt with truths whose full and complete comprehension lay outside of mathematics itself. Algebra had a noumenal aspect which lay within the province of morality and religion.. 'There is something mysterious and transcendent in the idea of Time'. (44) This meant that mathematics as a discipline was not a totally autonomous activity but stood in a subordinate relation to a higher form of knowledge. Like the case of inductive knowledge mentioned by Carlyle, it was only true <u>if</u> something else were true. And, as before, this doctrine about essences was simultaneously a doctrine about the relationships of institutions. Its practical import was to place mathematics as a profession in a relation of general subordination to the Church. Algebra, as Hamilton viewed it, would always be a reminder of, and a support for, a particular conception of the social order. It was symbolic of an 'organic' social order of the kind which found its expression in Coleridge's work on Church and State.

4. This approach makes it clear why Hamilton would be opposed to the formalism of the Cambridge school. Formalism would rob mathematical symbols of the particular social meaning he wanted them to carry. Instead of making algebra yield images

43. quoted by Hankins, 1976, p.342.
44. quoted by Hankins, 1976, p.350.

of hierarchy and provide glimpses of divinity, the formalist approach rudely collapsed hierarchies: there was here no hint of mystery or dependence on higher truths. It is therefore not surprising that when Hamilton described his reaction to Peacock's Algebra, he should convey this through language and imagery derived from lowly domestic objects.
Peacock, said Hamilton,

> designed to reduce algebra to a mere system of symbols, and nothing more; an affair of pothooks and hangers, of black strokes upon white paper, to be made according to a fixed but arbitrary set of rules: and I refused, in my mind, to give the high name of Science to the results of such a system. (45)

Formalist mathematics does not point upwards but downwards, to pothooks and hangers. If we leave out the reference to time, said Hamilton, mathematics would 'descend into the rank of an art' or turn into a language. 'The symbols will then become, what many now account them to be, the all-in-all of algebra'. (46) So Hamilton's intuitionism, and his rejection of formalism, was not idle speculation or a free-floating intellectual preference: it was indeed a practical doctrine. He was doing with mathematics what he and others were doing generally with the Idealist view of the world: he was interpreting it in a way that made it legitimate his social values.

What these social values were has already been glimpsed in Hamilton's reaction to the one-sided utilitarianism represented by Airy. His biographer fills out this picture in a way that links him even more strongly to Coleridge's brand of conservatism. Graves says that Hamilton 'at all times evinced a serious interest in political and social questions', though he insists that he was no partisan. (47)

45. Graves, vol. II, p.528.
46. Graves vol. III, p.634.
47. Graves, vol. II, p.556. (The point about Hamilton not being partisan is on p.100 of the same volume.)

What Graves means by not being partisan may perhaps be gathered from a letter that did not find its way into the biography. Writing to his M.P. on January 21st 1853, Hamilton donated a pound to the Conservative Registration Society of Dublin but added

> I do not wish my name to appear in print, in connection with that, or with any other political Society. (48)

Even so, Hamilton had publicly committed himself by joining the local conservative association as far back as 1834, and had actually addressed that body. (49) As he told De Morgan, in a letter which assumed that De Morgan would be in the opposite camp:

> Understand, first, that I don't pretend to be an unprejudiced man. Deeply prejudiced I know myself to be; not thereby admitting that I am wrong. From childhood I have had political leanings, and always on the <u>il</u>liberal side. (50)

Hamilton was naturally a loyalist and a patriot, emotionally involved with Britain's imperial adventures and anxious at her setbacks in colonial wars. (51) In 1848 he proclaimed his readiness to bear arms for the Queen in the event of rebellion. (52) He was a staunch Irish protestant with leanings towards the ritualism of the High Church (53), and accepting

48. <u>TCDMS</u> 1492/127.5.
49. Graves, vol. II, p.101.
50. Graves, vol. III, p.392.
51. Graves, vol. III, p.27. 'It was English history not Irish, which I was taught, and my head still throbs with sympathy for the great British Empire .... I was almost literally sick with sorrow at hearing of the disasters in Cabool and the Khyber Pass several years ago'.
52. Graves, vol. I, p.647. 'I have always been a loyalist, and was enrolled in the spring of 1848 among those who where ready to take up arms ...'.
53. Graves, vol. I, p.612. 'some people were pleased to call me a Puseyite, some years ago. However, I never pleaded guilty to the charge, though I had certainly leanings to high churchism'.

the ultimately mysterious character of religion. (54) His biographer asserts that he was

> habitually formal with a formality that sprang from a deep value for law in all things: he loved order and co-ordination and subordination and symmetry and completeness; and this love pervaded all his mathematical work. It was this love of order that made him in politics a large-minded Conservative, valuing liberty but valuing also subordination of ranks and supremacy of civil law; and that in matters of religion led him to recognise the importance of adding to individuality the outward organisation of an authoritatively constituted and graduated ministry ...... (55)

It was this 'love of order', and 'subordination of ranks', that aligned Hamilton with the interests served by Idealism, and which makes sense of his intuitionism. The isolation from his fellow scientists of which Hamilton complained was indeed a 'philosophical isolation' - as Hankins characterised it - but it was more than this: it was a political isolation. (56)

5. So far I have only looked at formalism through Hamilton's eyes. If the approach that I am taking is correct then it should illuminate formalism when set in its own context.

Both Hamilton and the Cambridge group were heirs to the same set of technical resources, and the same tradition of posing and solving problems. Both parties were very familiar with the use of imaginary quantities and the standard manipulations that gave rise to them, and shared the wide-spread dissatisfaction with the current understanding of these procedures. They both knew about Warren's geometrical representation of complex numbers, both apprehended its utility and both had reservations about it. Neither side was determined by these background considerations, as is shown by the fact

54. Graves, vol. I, p.465.
55. Graves, vol. I, p.451.
56. Hankins, 1976, p.340.

that at this point they diverged. Agreeing that Warren did not get to the heart of the matter, they disagreed on what to do about it.

For Peacock, Warren's work did not touch the essence of imaginary numbers at all. At most it represented a geometrical _interpretation_ of a set of symbolic manoeuvres. It was in the rules for manipulating the symbols that the real essence of imaginary numbers was to be found. In themselves imaginaries were just symbols like any others. What mathematicians had been doing all along - what they had been writing on the page - was sufficient in itself. All that was needed was a doctrine about the nature of algebra to make them aware of why this was so, and encourage them to proceed in the same way rather than getting sidetracked into geometrical justifications. All that the existence of a geometrical interpretation shows is that geometry can be brought within the scope of algebra. It simply extends its hegemony. Thus in his famous report to the British Association in 1833 Peacock had said

> The capacity, therefore, possessed by the signs of effection involving $\sqrt{-1}$ of admitting geometrical or other interpretations under certain circumstances , though it adds greatly to the power of bringing geometry and other sciences under the dominion of algebra, does not in any respect affect the general theory of their introduction or of their relation to other signs, for in the first place it is not an essential or necessary property of such signs ..... It would be a serious mistake, therefore, to suppose that such incidental properties of quantities affected by such signs constituted their real essence. (57)

57. G. Peacock, 'Report on the recent progress and present state of certain branches of analysis', _Report of the third meeting of the British association for the advancement of science held at Cambridge in 1833_, London, Murray, 1834, pp.185-352.

Peacock claimed not to be surprised that Warren's inverted order of priorities led 'to very embarrassing details' and to the neglect of 'comprehensive propositions'.(58) It seems to me, said Peacock

> to be a violation of propriety to make such interpretations which are conventional merely, and not necessary, the foundation of a most important symbolical truth. (59)

These remarks need to be seen in context. It had long been assumed in Cambridge, as elsewhere, that geometry was superior to algebra for its rigour, its clarity and pedagogic utility. Newton himself had written of the need to adorn algebra with geometry to make it elegant and 'fit for public view'. (60) Peacock was rudely and uncompromisingly inverting this order of priority. In doing this he was asserting and elaborating the stance that had defined the Cambridge group since 1811.

It is well known that in the Cambridge context the advocates of symbolical algebra were reformers and radicals. As undergraduates they founded the famous Analytical Society and published papers in its proceedings. They stood for a break with the past; for an end to traditional authority in mathematics teaching, and for the introduction of new techniques of analysis. (61) They took the discipline of mathematics seriously in the sense that they wrote for other mathemati-

58. Peacock, 1833, p.229.
59. Ibid., p.305.
60. Quoted by Baden Powell, 1830, p.478 from Newton's treatise on fluxions.
61. W. Rouse Ball, A history of the study of mathematics at Cambridge, Cambridge, C.U.P. 1889.
    J.Dubbey, The mathematical work of Charles Babbage Cambridge, C.U.P.1978, Ch.3.
    P. Enros, The Analytical Society : mathematics at Cambridge university in the early nineteenth century, (unpublished PhD thesis for the University of Toronto), 1979.

cians, lifting their eyes above the narrow horizon of collegiate requirements and loyalties. They were, in a sense, professionals. (62) They jokingly looked upon themselves as bringers of light into darkness, as reformers with 'extensive schemes for enlightening and improving the human race'. (63) But they missed no opportunities to put their ideas into practise. Peacock, for instance, exploited to the utmost his position as moderator in the Senate House examinations. (64)

Babbage told a revealing story about how the Analytical Society was founded. Allegedly this significant step in the introduction of new analytical methods was stimulated by the founding of the Cambridge branch of the British and Foreign Bible Society. Considerable controversy raged over whether the Bible should be distributed with or without an accompanying commentary. The Evangelical student group behind the new society objected to a commentary; the High Church Faction in the University favoured a commentary. The point was whether people could be trusted to reach acceptable conclusions from the Bible when left to their own devices. From the High Church point of view the Evangelicals were dissenters, and little better than the sectarian fanatics of old. What was wanting in their theology was a decent respect for _obedience_. (65) Babbage reports that on hearing about this

62. Enros argues that the Analytical Society was not founded with the aim of reform as such, but rather with the aim of allowing its members to contribute to 'professional' mathematics, that is, mathematics produced by an internationally based group of specialists devoted to abstract, disciplinary criteria. It was the perceived _failure_ of this venture that translated itself into reforming zeal.
63. Letter from Babbage to Herschel, 1st Aug. 1814.
64. See Enros, p.234.
65. Ford Brown, _Fathers of the Victorians: the age of Wilberforce,_ Cambridge, C.U.P. 1961, p.299.

issue his mind leapt to the idea of a society for distributing copies of Lacroix - one of the French textbooks of analysis. It would be a shining example of the truth which everyone could see for themselves by the light of the reason. The new society, Babbage decided, would maintain 'that the work of Lacroix was so perfect that any comment was unnecessary'. (66)

The same themes of reform, change and autonomy are found if we look at the broader spectrum of attitudes that were adopted by the formalists of the Cambridge group. Babbage is perhaps an extreme case, but he certainly threw himself vigorously into the world of liberal politics and electioneering; of science in the service of industry, where the scientist could adopt the role of consultant and expert. Even his theology implies that science is an activity independent of any higher spiritual authority. This comes over in his theory of miracles which he explains as singularities in the equations which describe natural processes. Peculiar events which might suggest divine intervention are really part of the self-contained operation of the order of nature. In sharp contrast to Hamilton's account we read that

> Miracles, therefore, are not the breach of established laws, but they are the very circumstances that indicate the existence of far higher laws, which at the appointed time produce their pre-intended results. (67)

66. C. Babbage, _Passages from the life of a philosopher_, London, Longman, 1864, p.28. (When Lacroix was finally translated as part of the reform programme it is interesting to note that it _did_ have a commentary. This explained that Lacroix was wrong to use the method of limits but should have developed the calculus on the basis of infinite series after the fashion of Lagrange.)
67. _Ibid_. p.391.

And what Babbage meant by higher laws was explained by reference to his famous calculating engine which could be made to produce analogues of 'miracles' as it switched from one programme to another in a predetermined way. (68) So God is brought down to earth and included in the scheme of things with scientists - and scientists alone - can understand. Babbage's oft-quoted witticism about the Cambridge group advocating 'Pure D-ism' was perhaps more than a pun about their mathematical notation. (69)

De Morgan was not one of the founders of the symbolical school but, after leaving Cambridge on religious grounds, he carried on their work as professor at the new University College, London. Set up as a middle class rival to Oxford and Cambridge, University College was 'as much a political as an educational challenge, threatening the secure elitism of the Tory landed and clerical establishment by its unabashed business character ..... and even more by its disturbing secularism'. (70) De Morgan's individual religious commitment was accompanied by a fierce concern to keep the College true to its principles of being free from orthodox religious tests and constraints. His demand for doctrinal purity on this question, and the absolute autonomy of the professoriate, was extreme and uncompromising, and his professional life was marked by battles over this question. (71)

68. _Ibid._, Ch.XXIX.
69. _Ibid._, p.29.
70. Leslie M. Crossley, _The professionalisation of science in Victorian Britain_, unpublished PhD thesis, Univ. of New South Wales, 1979, p.95, and H.H. Bellet, _University College London 1826-1926_, London, Univ. of London Press, 1929, Chapters I and II. For Coleridge, University College was just a 'lecture bazaar', _Ibid._, p.80.
71. Sophia De Morgan, _Memoir of Augustus De Morgan_, London, Longmans, 1882, sections II and XI.

As well as his work for insurance companies De Morgan was a contributor - indeed a massive contributor - to the publications of the Society for the Diffusion of Useful Knowledge. (72) This symbol of self-help and independence also had the allegiance of Hamilton's friend John Graves who sided with the formalists. Hamilton himself declined Lord Brougham's invitation to become involved in the Society. (73)

De Morgan had family links with some of the political radicals of the previous generation. He was married to a daughter of William Frend, who had been banished from Cambridge for his republican sentiments and had known Coleridge in his early radical phase. (74) Unlike the poet, Frend had stayed a supporter - albeit an increasingly eccentric one - of progressive and revolutionary causes. Frend had been an able mathematician and had published a textbook which De Morgan declared to be 'on the points which it treats, the clearest book in our language'. (75) What made him unusual, however, was that he totally rejected the concept of negative or imaginary magnitudes. De Morgan did not agree with this stance but his analysis of its causes is itself revealing.

72. The memoir gives a list, over six pages long, of articles written by De Morgan for the Penny Cyclopaedia. The list begins on p.407.
73. For John Graves' attachment to the symbolical school see Graves, vol.II, p.143; for his contributions to the S.D.U.K. see vol.II, p.378; for Hamilton's refusal see vol.II, p.127.
74. Coleridge as an undergraduate attended Frend's trial at the University, Frend being a fellow and tutor at Jesus. See John Colmer, _Coleridge, critic of society_, Oxford, Clarendon, 1959, p.4.
75. A. De Morgan, 'Report of the council of the society to the Twenty-Second annual general meeting', _Monthly notices of the Royal Astronomical Society_, vol.5, 1842, p. 150.

Frend adopted his position, De Morgan tells us in his long obituary, precisely because of the social and political meaning that had become attached to these obscure concepts. Because the meaning of negatives and imaginaries was mysterious they were taken up and used to justify the mysteries of religion, and these mysteries in their turn were associated with the reactionary and repressive machinery of Church and State that hounded Frend for his republicanism. (76) De Morgan goes on to add that a proper (formalist) understanding of negatives and imagineries is now possible, so that Frend's self-denying strategy is obsolete, although he adds, the practice of using these mathematical ideas to bolster religion is

> not extinct in our own day, even after all that was inexplicable about impossible quantities has disappeared. (77)

In general the Cambridge group were liberal rather than Tory. This can even be said of the young Whewell. Todhunter tells us that

> In politics Dr. Whewell in early life was not altogether averse from the Whig party, which included some of his firmest friends, as Sheepshanks and Peacock; but he became in the end a Conservative .... (78)

It is also significant that the young - though not the old - Whewell was scathing about the Lake poets. Wordsworth's 'Excursion', Whewell noted in 1820, was a prolix and feeble attack on the manufacturing system; Coleridge's 'Lay Sermons' were a 'most impudent attempt to make reason commit suicide'; while Southey was just a tory 'bigot'. (79)

76. Ibid. p.146.
77. Ibid. p.146.
78. Todhunter, 1876, vol.I, p.413.
79. Ibid., vol.I, p.356.

Why, then, did the Cambridge mathematicians such as Peacock, Babbage, the young Whewell, De Morgan and the other followers of the analytical movement treat symbols as 'the all in all' of algebra? Why did it seem natural to treat symbols as self-sufficient, as if mathematics were essentially about marks on paper? Although the evidence is scattered, enough has been said to suggest that here too the answer lies in the social meaning and use that was attached to the doctrine. The mathematicians imputed self-sufficiency to their symbols when they, their users, were asserting their own self-sufficiency and impressing that fact on others. Formalism was useful to the emerging 'professional' mathematicians of Cambridge and London because it brought mathematics entirely within their grasp. It made it out to be an internal system of meanings in which no one else had a legitimate interest. It celebrated the self-sufficient character of mathematics, and hence the self-sufficient character of mathematicians. (indeed, formalism might almost be called mathematical pantheism.) Conversely, symbols were denied autonomy and were portrayed as standing in need of reference to something ideal when their users - like Hamilton - wanted to impress on others the need for an analogous dependence in the social realm. Stated in its broadest terms, to be a formalist was to say: 'we can take charge of ourselves'. To reject formalism was to reject this message. These doctrines were, therefore, ways of rejecting or endorsing the established institutions of social control and spiritual guidance, and the established hierarchy of learned professions and intellectual callings. Attitudes towards symbols were themselves symbolic, and the messages they carried were about the autonomy and dependence of the groups which adopted them.

6. I do not pretend that this account is without problems or complicating factors. For example, it would make the story simpler if Hamilton had been an out and out reactionary like his other poet-mentor, Wordsworth. The fact is that, in politics, unlike Coleridge and Wordsworth, Hamilton was suffi-

ciently moderate - or sufficiently subtle - to support the Reform Bill of 1832. In their correspondence Wordsworth put considerable pressure on Hamilton to define his position on political questions, and in a letter dated June 13, 1831, reported

> I saw little or nothing of Cambridge on my return - which was upon the eve of the election - but I found that the Mathematicians of Trinity, Peacock, Airy, Whewell, were taking what I thought the wrong side .... Your University, I am proud to see, keep to members that do it credit. (80)

In embarrassed sentences, of a complexity remarkable even for Hamilton, he feels forced to 'confess' to Wordsworth that he too supports reform. (81) But this need not, I think, cause too much trouble for my thesis. There were many tactical reasons for supporting reform. As Hamilton said to his sister: 'I am a reformer chiefly because I prefer gradual to a sudden revolution'. (82)

Again it is necessary to notice and account for the fact that Hamilton's opposition to Cambridge formalism seemed to decline with time. In a letter to Peacock dated Oct. 13, 1846, Hamilton declared that his view about the importance of symbolical science 'may have approximated gradually to yours'. (83) Interestingly, Hamilton also noted some four years later 'how much the course of time has worn away my political eagerness'. (84) A corresponding and opposite movement took place in Whewell's life. Here, in obliging conformity with my thesis, it is known that as Whewell moved to the right, and embraced the ideology of liberal education rather than the more esoteric ideals of the Analytical Society, he increasingly moved away from the symbolical approach in his

80. Graves, vol.I, p.428.
81. Graves, vol.I, p.478.
82. Graves, vol.I, p.537.
83. Graves, vol.II, p.527.
84. Graves, vol.II, p.653.

mathematical writings. (85)

Perhaps the most important shortcoming in the present account concerns the evidence that I have adduced in my attempt to contextualise Hamilton and Peacock. To make my conclusions secure and the case persuasive, it would be necessary to locate Hamilton much more securely in his Dublin surroundings. I need to know with much greater precision what it meant to be an Irish tory, or a Coleridgean, or a professor (but not a fellow) of Trinity College Dublin. (86) I have not been able to relate Hamilton's Idealism to his immediate institutional surroundings or to find for it any context of employment other than the rather diffuse arena of national politics. As far as the Cambridge group is concerned, similar doubts may be raised. Perhaps all that needs to be said about their formalism relates to the conflicting interest within the University of Cambridge (or in De Morgan's case, University College), and I may have weakened the argument by introducing broader political alignments.

Despite these shortcomings, and with all its defects as history, my argument at least serves to illustrate a significant explanatory principle and to show how it may be applied to mathematics. The principle is that in our social life we are always putting pressure on our fellows and seeking to evade that pressure ourselves - trying to preserve a customary pattern of behaviour or an established institution, or trying to change them to our advantage. In order to do these things we try to make reality our ally, showing how the nature of things supports the status quo, or how the establish-

85. For a valuable discussion of the changes in Whewell's position see Enros, pp. 247-255.
86. The potential importance of this distinction was pointed out to me by Gordon Herries-Davies of Trinity College, Dublin.

ed social order is at odds with what is natural. This is why nature is so frequently populated with causes and principles which will punish the wicked and reward the good. It is why illness will follow immorality or disaster result from disloyalty. More generally, this is why we so often make the pattern of nature reflect the pattern of the social arrangements we desire. All this derives from nature being given a social employment. There is ample evidence to show that nature frequently is given such employment, and that this can explain much that is believed about it. Anthropologists are familiar with this phenomenon, (87) and recently historians of science have shown its operation in the development of the theory of matter. (88) The case of Hamilton and Peacock shows how this argument may be extended into mathematical reality. The rival theories of the essence of algebra that we have examined each seemed to carry a social message and this may be sufficient to explain their location and their differential credibility.

87. The point has been stressed by Mary Douglas. See esp. Mary Douglas, Purity and danger: an analysis of concepts of pollution and taboo, London, Routledge & Kegan Paul, 1966, Ch.5; Mary Douglas, Natural symbols: explorations in cosmology, Harmondsworth, Penguin, 1973; Mary Douglas, Implicit meanings: essays in anthropology, London, Routledge & Kegan Paul, 1975, Part 3.
88. See the references to the work of J.R. Jacob and M.C. Jacob in footnote (30).

Acknowledgements

I am greatly indebted to the participants in the Berlin conference for their constructive scepticism about the first draft of this paper which was little more than the bare bones of an idea, viz. that perhaps the work done by Jacobs on the theory of matter may be extended to cover mathematical symbols as well (see footnote 30). The most that I have been able to do is to produce a little more by way of illustrative material to flesh out this idea. Outside the conference room I am particularly indebted to Professors Michael Crowe and Elizabeth Garber, and to my colleagues Steven Shapin, David Miller and Stephen Jacyna for gently but firmly making clear to me how far the present paper is from being an acceptable piece of historical scholarship. The point is well taken: the most that I dare claim is that it may have some interest as an illustration of what a certain type of explanation may look like when applied to mathematicians. I must also record my debt to the Royal Society of London who enabled me to examine the Hamilton archives in Dublin.

THE BERLIN SCHOOL OF MATHEMATICS

Thomas Hawkins

Anyone at this workshop who happens to be familiar with my publications in history of mathematics would probably classify me as a practitioner of what Mehrtens and others call the internal history of science. Although I would not dispute such a categorization - and I am delighted that Mehrtens sees a future for internal history - I would be the first to admit that internal history in the strict sense is insufficient to provide a complete understanding of the growth of mathematical knowledge. Historical studies that fall under the rubric of social history of mathematics are essential to our understanding of the history of mathematics. In his paper Mehrtens has indicated a number of diverse approaches to history that might constitute social history of mathematics, especially 19th-century mathematics. I agree wholeheartedly with him that it would not be wise to discourage diverse modes of research in favor of a "unified" approach to social history. Indeed I would like to suggest a further way in which social history might be pursued to the benefit of, and in conjunction with, internal history.

In order to understand the growth of mathematical knowledge it is sometimes important to identify and consider the role played by schools of mathematical thought. Such a school usually possesses an underlying philosophy by which I mean a set of attitudes towards mathematics. The members of a school tend to share common views on what kind of mathematics is worth pursing or, more generally, on the manner in which, or the spirit in which, one should investigate mathematical problems. The effects of this philosophy upon mathematicians associated with a school must be taken into consideration by the internalist historian seeking to comprehend the elements providing the dynamics of growth of mathematical knowledge. He must thus be conscious of and sensitive to the social context in which mathematics is created or within which a mathematician received his training. Unfortunately relatively little seems to have been done along the lines of seeking to identify such schools, to clarify the nature of their philosophies and to study their effect upon the development of mathematics. Such an undertaking seems to me an ideal meeting ground for historians of diverse methodological persuasions and one, in particular, that requires work in social history of mathematics. There is also the further question, not to be concidered here, of how the philosophies of various schools are themselves the product of social and internal factors.

My own appreciation of the significance of underlying schools of thought has increased over the past few years by virtue of my researches on the work of Georg Frobenius and

Wilhelm Killing, two mathematicians associated with the Berlin school of mathematics centered about Weierstrass. The work of Frobenius that interested me had to do with matrix algebra. In the case of Killing I was concerned with his fundamental contributions to the structure of Lie algebras which grew out of his involvement with the foundations of geometry in the light of the discoveries in non-Euclidian geometry. In both cases, therefore, I was concerned with work in areas apparently far removed from the Weierstrassian analysis with which we usually associate the Berlin school. I discovered that, nonetheless, both Frobenius and Killing were guided in their work by the Weierstrassian philosophy that they had acquired during their training at Berlin. (Neither was in Berlin when the relevant work was actually carried out.) In order to fully grasp the motivation behind their work and why it took the directions it did, it was necessary to be mindful of the school that had produced them. In the remainder of this paper I shall briefly sketch the manner in which the philosophy of the Berlin school motivated and informed the mathematics of Frobenius and Killing. Although in both cases, I shall be sketchy, I shall be even more so in my treatment of Frobenius since the details can be found in my publications. I would also like to indicate my indebtedness to Kurt R. Biermann for his informative study, _Die Mathematik und ihre Dozenten an der Berliner Universität 1810-1920_ (1973), which greatly facilitated my own work. A similar study of the Göttingen of Hilbert and Klein would be equally useful and important.

Both Frobenius and Killing began their studies at the University of Berlin in 1867, obtained their doctorates under the direction of Weierstrass, and then remained in Berlin until 1875 and 1877, respectively, earning their livings at various teaching positions. They do not appear to have been close friends, although both shared the experience of being a part of the circle of students about Weierstrass. Neither of them, however, was primarily interested in mathematical analysis. Killing was a geometer at heart, while Frobenius' first love was algebra and the theory of numbers. Yet each made important contributions to their preferred areas of mathematics that were inspired by the spirit of Weierstrassian mathematics, particularly (although certainly not exclusively in the case of Killing) as expressed in his theory of elementary divisors (1868). This theory represents Weierstrass' critical response to the contemporary practice of finite or algebraic analysis and is less wellknown than his response to the practice of infinitesimal analysis.

Ever since its inception, the method of symbolical analysis was characterized by its generality, the source of its extraordinary power. This led to a tendency on the part of analysts to reason on what I have termed the generic level. In analytical reasoning one is dealing with symbolical expressions, and the analyst tended to regard the symbols as taking "general" rather than particular values, to deal, with varying degrees of awareness, with the "general" case, thereby tending to ignore the singular types of situations that could occur for certain specific value assignments of the variables. One particular area of algebraic analysis dominated by generic reasoning was the theory of the transformation of quadratic and bilinear forms by linear substitutions. Weierstrass' theory of elementary divisors is devoted to a rigorous exposition of this theory, i.e. one that is not carried out on the generic level but satisfies the Weierstrassian demand for an analysis that covers all the "special cases" that can possibly arise.

The theory of elementary divisors thus became an integral part of the Berlin philosophy. To his colleagues and students at Berlin, Weierstrass had demonstrated more than theorems in his paper on quadratic and bilinear forms: He had also demonstrated the desirability and feasibility of a more rigorous approach to algebraic analysis, an approach that did not rest content with the prevailing tendency to reason in terms of the "general" case. Kronecker expressed the credo aptly in one of his many papers (1874) inspired by that of Weierstrass:

> It is common - especially in algebraic questions - to encounter essentially new difficulties when one breaks away from those cases which are customarily designated as general. As soon as one penetrates beneath the surface of the so-called singularities, the real difficulties of the investigation are usually first encountered but, at the same time, also the wealth of new viewpoints and phenomena contained in its depths.

These sentiments were a driving force behind the work of Frobenius and Killing.

In the period 1878-80 Frobenius published several memoirs of great significance for the history of the theory of matrices. In them he introduced, independently of Cayley's little known memoir of 1858, the symbolical algebra of matrices (or forms as he called them), developed its consequences far beyond the level of Cayley's work, and demonstrated the advantages attained by fusing matrix algebra with the theory of canonical matrix forms (as per Weierstrass' theory of elementary divisors). With these papers the theory of matrices as we now conceive it commenced its existence. Frobenius' mathematical activity along these lines was initiated by a certain problem I have termed the Cayley-Hermite problem. It is the problem of determing the linear substitutions which leave a given non-singular quadratic form invariant. Why did this problem attract Frobenius' interest? The reason is that its study by Hermite, Cayley and a few others had been carried out on the generic level. Here, then, was a good Weierstrassian problem. As Frobenius himself put it in his first paper: "Investigations of the transformation of quadratic forms into themselves have so far been limited to consideration of the general case, while exceptions to which the results are subject in certain cases have been exhaustively treated only for ternary forms [by Bachmann (1873), who obtained his doctorate from Berlin under Kummer].... I have thus attempted to fill in the gaps which occur in the proofs of the formulas ...." In order to fill in the gaps in a manner that was at once rigorous and elegant, Frobenius devised his symbolical algebra of forms and fused it with the canonical matrix forms of the theory of elementary divisors of Weierstrass and Kronecker.

The role of the philosophy of Weierstrassian mathematics is even more interesting in the case of Killing because the points of contact with his work are more varied. Wilhelm Killing began his studies in Berlin in the Winter Semester of 1867. He had already spent two disappointing years at the University in Münster, not far from his home, where he was forced to learn mathematics on his own because Münster offered only elementary mathematics taught by an observational astronomer. By virtue of the extreme contrast, Berlin was all the more impressive to Killing. At Berlin, Killing was attracted above all to Weierstrass, who by Killing's own admission exerted the greatest influence on his scientific education. Although he was greatly impressed by the theoretical emphasis of Weierstrass' lectures, Killing's principal mathematical interests, since his youth, lay in geometry, not analysis. While at Münster he had made a careful study of the treatises on analytical geometry by Hesse and Plücker. The standard treatment of quadric surfaces, when viewed in the critical spirit of Weierstrass, appeared inadequate and Killing set himself the task of providing an exhaustive, Weierstrassian analysis of all the geometrical possibilities, however uninteresting they might be from a geometrical viewpoint, for a pencil of quadric surfaces. His main mathematical tool was the theory of elementary divisors, and in fact Killing described his dissertation as a geometrical interpretation of Weierstrass' theory. Indeed Killing's dissertation could be described more generally as a geometrical interpretation of how to pursue geometry in the spirit of the Weierstrassian philosophy. By emphasizing the need to consider all the geometrical possibilities revealed by analysis, regardless of their intrinsic geometrical significance, Killing was advocating an approach to geometrical problems that was not typical of the period. The same characteristics are to be found in his work on foundations of geometry, which was even more untypical of contemporary work than his dissertation had been.

Killing's interest in the foundations of geometry seems to have begun with some lectures on the subject by Weierstrass in the Mathematics Seminar for the Summer Semester of 1872. By that time foundations of geometry was inseparable from the issues raised by discoveries made in non-Euclidian geometry. Let me recount those discoveries as regarded from the perspective of the Weierstrassian philosophy. Non-Euclidian geometry began with the discovery of Lobachevskian geometry which showed that Euclidian geometry is not the sole geometry that is logically consistent and compatible with experience, insofar as that compatibility can presently be measured. Lobachevskian geometry involves a certain parameter k, which may be regarded as a radius of curvature and which for $k = \infty$ yields Euclidean geometry as a limiting case. Tha value of k, however, cannot be determined _a priori_ so that geometry becomes an empirical science. The discoverers of Lobachevskian geometry tended to regard it as an "absolute geometry". That is, Lobachevskian geometry, including thereby the limiting case of Euclidian geometry, was tacitly regarded as the only conceivable geometry. But then Riemann made the observation that the unboundedness of space is more empirically certain than its infinitude. He also observed that Gauss' intrinsic differential geometry of surfaces could be generalized to higher dimensions so that space could be conceived of as a manifold with curvature. Within this conceptual framework it was then possible to distinguish unboundedness from infinitude and to realize a space that is finite yet unbounded as a manifold of constant, positive curvature. Although Riemann's discussion of these manifolds is vague, it would seem that he identified the geometry of such a manifold with spherical geometry. In any case his readers made such an identification. It was thus a further revelation when Felix Klein and Simon Newcomb independently discovered a different geometry for a manifold of constant positive curvature, which Klein named elliptic geometry.

Each new discovery revealed the unjustified limitations of the prevailing perception of the geometrical possibilities. The situation was analogous to that in the theory of functions of a real variable. The discovery of new types of functions, such as continuous nowhere differentiable functions, revealed the untenable nature of the prevailing perceptions of functions and their properties. Furthermore the attempted proofs that the above three geometries - or in some instances just Lobachevskian and spherical geometry - were the only possible non-Euclidian geometries, made tacit use of "facts" derived from intuition but not specifically posited, just as did the proofs that a continuous function must possess a derivative at "most" points. The mathematicians who accepted the Weierstrassian philosophy rejected such proofs and accepted the existence of counter-intuitive functions and the need to deal with them systematically and rigorously. In the realm of geometry Killing stood as the solitary exponent of an analogous attitude towards research in the foundations of geometry. (He even used the existence of continuous nowhere differentiable functions to refute one geometer's proof that aside from Euclidean geometry only spherical and Lobachevskian geometry are possible.) Killing therefore proposed a systematic and exhaustive analysis of all possible forms of space (*Raumformen*) which would not in any way rely upon intuition and which, he stressed, would involve consideration of space-forms that directly contradict our intuitions and experience of space.

In order to carry out the program of determining all space-forms, Killing followed the lead of his teacher at Berlin, Hermann von Helmholtz, who (with quite different intentions!) had approached the problem of deducing the properties of space from the behavior of the motion of rigid bodies. In fact, in his attempt to derive the quadratic nature of infinitesimal distances, ds, which Riemann more or less hypothesized, Helmholtz actually worked exclusively with the infinitesimal motions of space. Killing followed suit. The problem of determining all n-dimensional space-forms therefore translated into the problem of determining all the infinitesimal motions of space-forms with "m degrees of mobility". Without going into any technical details, let me just state that the infinitesimal motions with m degrees of mobility form what is now called an m-dimensional Lie algebra. Hence one component of Killing's research program on the determination of all space-forms was the determination of the structure of all Lie algebras. Killing's contributions to the solution of this formidable problem are still fundamental to the theory of the structure of Lie algebras today. Indeed, Killing's work really initiated the theory, as well as the structure theory of rings and algebras. The important point I wish to make here is that the peculiar nature of Killing's research program in foundations of geometry was a reflection of the Berlin philosophy as applied by Killing to the domain of geometry. He sought to exhaustively analyze space-forms just as he had previously done for pencils of quadric surfaces in his dissertation under Weierstrass. Furthermore, it was precisely the untypical orientation of his work that led him to formulate, via the problem of determining all possible space-forms, the problem of determining all possible finite-dimensional Lie algebras. Thus by paying attention to the philosophical orientation of a mathematician, one can be led to unexpected connections, such as the role of the Berlin school of Weierstrass in the history of Lie algebras.

In order to add another dimension to my arguments for the role of the Berlin School in Killing's work, I also had occasion to consider two other schools of thought, namely those of Plücker (Bonn) and Clebsch (Göttingen), which shared an antipathy towards the Berlin school. These were the schools in which Felix Klein was trained. In order to bring out more fully the Weierstrassian character of Killing's work I deemed it instructive to consider the work on non-Euclidean geometry of another mathematician who was not influenced by Weierstrass, who in fact was consciously opposed to the spirit in which mathematics was approached at Berlin. Klein was such a mathematician. He is particularly interesting in this connection because there are many superficial similarities between his papers on non-Euclidean geometry and the ensuing Erlanger Programm on the one hand, and Killing's theory of space-forms on the other - so much so that Killing felt that Klein's "projective space forms" required nothing more than a natural and "slight" generalization in order to become his own space-forms. By carefully analyzing Klein's work on, and attitude towards, geometry, I conclude that Killing could not have been more mistaken regarding the affinity of their respective geometrical studies. In this connection, consideration of the schools that produced Klein is helpful and enlightening, for their tenets can be seen reflected in Klein's attitude. Klein of course was a natural leader and eventually established his own school at Göttingen. In his lectures on non-Euclidean geometry at Göttingen (1891-2) Klein contrasted the philosophies of Berlin and Göttingen:

> With what should the mathematician concern himself. Some say: certainly intuition is of no value whatsoever; I therefore restrict myself to the pure forms [Gebilde] generated within myself, unhampered by reality. That is the password in some places in Berlin. By contrast, in Göttingen the connection of pure mathematics with spatial intuition and applied problems was always maintained and the true foundations of mathematical research recognized in a suitable union of theory and practice.

The approach to mathematics which Klein so unsympathetically portrays was, as Klein sensed, a part of the Berlin ambience. The contrasts he drew between Berlin and Göttingen apply in particular to Killing's and his own approach to geometry.

Bibliographical Remarks

The details connected with my discussion of Frobenius can be found in my papers: "Another Look at Cayley and the Theory of Matrices", Arch. int. d'hist. sci., 26 (1977), 82-112; "Weierstrass and the Theory of Matrices", Arch. Hist. Exact. Sci., 17 (1977), 119-163. The latter also contains an historical analysis of the origins of the theory of elementary divisors. The above discussion of Killing is based upon my essay, "Non-Euclidean Geometry and Weierstrassion Mathematics: The Background to Killing's Work on Lie Algebras", Historia Mathematica 7 (1980). Killing's actual contributions to the structure and representation of Lie algebras are treated in papers currently being prepared for publication.

F. SCHLEIERMACHER'S INFLUENCE ON H. GRASSMANN'S MATHEMATICS

Albert C. Lewis

In this paper I would like to summarize my account of the nature of the influence of Schleiermacher on Grassmann's mathematics and to make some comparisons between this and the role of W. R. Hamilton's metaphysics in Hamilton's quaternions. There are two questions that might be asked about Schleiermacher's influence: 1) How did Schleiermacher influence Grassmann's mathematics? and 2) Did this influence affect the acceptance of Grassmann's work in the mathematical community and, if so, how? Historians of other aspects of nineteenth-century mathematics may have an interest in the matter of the lack of contemporary acceptance of Grassmann's Ausdehnungslehre of 1844 since the reasons for this would say something about the mathematical community of his time. Parallels might be drawn with the reception given to Bolyai, Bolzano, Galois, and even Gauss (at least in his anticipation of a negative reception which kept him from publishing on non-Euclidean geometry). As D. Struik pictures it for us in this volume, it was the best of times in many ways for new, revolutionary, ideas in mathematics, while also the worst of times from the point of view of the reception, propagation and growth of some of these ideas. If Grassmann was trying to provide a sound, "wissenschaftlich," basis for his new branch of mathematics by tying it strongly to his father's views on foundations of mathematics and to Schleiermacher's Dialektik (as is maintained in (Lewis 1975 and 1977)) then why did he not explicitly state that this was his intention? On the one hand he comes close in the Ausdehnungslehre of 1844 to founding a philosophy of mathematics as well as a branch of mathematics, but, on the other, explicitly states that he is restraining himself from giving a full philosophical discussion because it will tend to turn mathematicians away:

> Es herrscht nämlich noch immer unter den Mathematikern und zum Theil nicht mit Unrecht eine gewisse Scheu vor philosophischen Erörterungen mathematischer und physikalischer Gegenstände; und in der That leiden die meisten Untersuchungen dieser Art, wie sie namentlich von Hegel und seiner Schule geführt sind, an einer Unklarheit und Willkühr, welche alle Frucht solcher Untersuchungen vernichtet. (Grassmann, 1844)

I will not, however, be able to take up the matter of this "extended" influence of Schleiermacher, i.e., the influence on the reception of Grassmann's work, but hope that further investigations will help shed light on it.

Grassmann

Hermann Grassmann's father went through university with the intention of entering the ministry. Hermann had two uncles, one of whom became a schoolmaster, the other a minister. He also prepared himself and took examinations to qualify for both the ministry and teaching. He took his first theological examination in 1834 and the same year succeeded Jakob Steiner as mathematics teacher at the Berliner Gewerbeschule. Immediately after taking the second theological examination in 1839 he applied for the further examination in teaching (Erweiterungsprüfung) in mathematics, physics, chemistry, and mineralogy. The theme, the theory of tides, provided the first opportunity for presenting those mathematical ideas which evolved into the calculus of extension of the Ausdehnungslehre of 1844. He continued teaching at secondary schools the rest of his career, eventually succeeding his father at the Gymnasium in Stettin. His interest in philology was expressed from an early age as well and he wrote a number of school texts for Latin and German. His translation and dictionary of the Rig-Veda is still considered a standard work in the field. The Ausdehnungslehre probably represents Grassmann's best expression of the interrelationship of his interests in language, pedagogy, theology, and mathematics.

Grassmann's philosophy is not explicit in the Ausdehnungslehre but I believe it can be explained by reference to Schleiermacher, especially to his Dialektik. The Dialektik might be described as Schleiermacher's counter response to Kant's transcendental philosophy. Originally presented as lectures in Berlin in the 1820s, attended by Grassmann, the Dialektik was published in 1839 and read by Grassmann in 1840. It represents a mature expression of Schleiermacher's early work Grundlinien einer Kritik der bisherigen Sittenlehre (1803). Corresponding to Grassmann's emphasis on what modern observers would regard as "externals" to mathematics -- method of presentation and education -- is Schleiermacher's emphasis on mathematics as resulting from creations by individuals and the implication from this that mathematics is a social creation as well as a collection of knowledge. This is an example of a typical Grassmann-Schleiermacher dilemma.

On the one hand, mathematics appears dependent on individual, idiosyncratic creativity (with all of the social influences this implies through the individual) and, on the other, it has to have also independence from individual creators in order to have its own abstract logical requirements.

Grassmann is credited with being the first to make an explicit break between mathematics as an abstract science whose objects of study, according to Grassmann, are created by thought, and mathematics as applied in the study of physical space and time. His program as outlined in the Introduction to the Ausdehnungslehre is far-reaching -- it is an effort not just to replace geometry (as then conceived as the science of physical space) with the abstract science of the calculus of extension, but also to present a unified basis for all the branches of mathematics -- number theory, combinatorial analysis, algebra, and theory of functions. But some historians in an effort to look for antecedents of the modern abstract, axiomatic, view of mathematics have overlooked the dialectical balance Grassmann considered necessary between abstract results and concrete instances of creation and learning -- both a part of mathematics as a science.

For Grassmann the dialectical dilemma referred to is not to be resolved away, rather it is a proper reflection of the true nature of mathematics. This is a characteristic of the polar contrasts in the Grassmann-Schleiermacher dialectical philosophy. Such contrasts are characterized by the following: (i) the One and the Many appear in some aspect in each contrast, for example, general and particular, continuous and discrete, and equal and different; (ii) relativism -- each of two opposite qualities depends on the other for its definition and is not to be thought of as describing a pure existent which has this quality alone; (iii) non-resolution -- the essence of reality is represented by the tension between contrasting elements rather than by their synthesis or resolution; and (iv) the contrasts are used as the determinants of concepts and of the species-genus relationships between those concepts.

These contrasts are reflected in every facet of the Ausdehnungslehre. The primitive mathematical entities are generated by the basic contrasts of equal-different and continuous-discrete; differences in generation are reflected by different steps of entities and by cognate steps of connections; and within each type of step there are opposites -- entities of the same and opposite senses of generation, and synthetic and analytic connnections.

According to Grassmann, mathematics proceeds from the particular to the general, and in the Ausdehnungslehre this movement is seen in the interplay between real and formal. Each mathematical concept may be said to have a real and formal aspect, and the relation of these two aspects has to be established in order for the concept to be completed. This is not to say such concepts are in fact completed in any absolute sense -- completeness appears very much relative for Grassmann as it is in the Dialektik. Concepts in the Ausdehnungslehre are at most complete within the theory of extension -- another context would presumably require a different 'determination of concept'. As with contrasts in the Dialektik, the real-formal and particular-general contrasts can be viewed as expressions of a many-one contrast. For example, different types of multiplication appear in all four branches of mathematics (presumably -- besides the theory of extension, Grassmann mentions only arithmetic and, implicitly, through reference to his father Justus, combinatorial analysis) and the properties common to all are given in the general theory of forms. Outer multiplication begins essentially with a real foundation or source in the particular branch, the theory of extension, and in stages those properties are developed which bring it under the formal category of multiplication. (Later in the Ausdehnungslehre another multiplication, the regressive, is introduced which is divided into two types, real and formal.) However, it should be noted that generalization is not the goal of mathematics but rather one pole of a contrast which governs its method.

Schleiermacher

According to most commentators Schleiermacher's distinct contribution to modern Protestant theology is the emphasis on the idea of Christ as a historically real person with a personality to be encountered by the Christian as his central act as a Christian. It is possible to see aspects of this theme in other ways, for example, in his strong admiration of Plato's dialogues where progress is made by a kind of dialectical process of one person directly confronting another. Also, in his praise of the "heuristic method" and even in his few remarks about history of science and mathematics, we can see an emphasis on personal invention which goes into the creation of science.

Of course, Schleiermacher's main concern was religion and in The Christian Faith he developed a formal and abstract religion. But alongside this he stressed that religion always appears in the world in a particular social and historical form. Analogous statements are made about mathematics by Schleiermacher and Grassmann. Thus, just as Schleiermacher stressed the centrality of Jesus Christ in Christianity and the need for the Christian to relate to the founder of Christianity, so Schleiermacher, and especially Grassmann, stressed the centrality of the mathematician in the particular historical development of mathematics and the need of the person (student or mathematician) studying mathematics to take into account how it was created since this is the best way of truly understanding the mathematics.

Schleiermacher's Dialektik is probably his most abstract work in that it concerns his search for a description of that which makes science possible. How, for example, an impersonal science can develop out of the personal contributions of individuals; how there can be any unity to a disparate range of sciences that somehow go to make up what we call scientific knowlege; and what the essential distinction is between the exact sciences of physics and mathematics and such knowledge as makes up ethics and theology.

Grassmann and Hamilton

The many similarities in the lives and works of Grassmann and Hamilton are brought out in Crowe's book (1967). Since the notion of generation or succession as the most fundamental mathematical idea appeared in J. F. Herbart and has a rather obvious similarity to Naturphilosophie ideas, it is not surprising to find it also, albeit in different forms, in Grassmann's generation of entities and Hamilton's succession of epochs of time. This appears to be one of those ideas (along with non-Euclidean geometry and the need to reform algebra, for example) which were 'in the air' during the first several decades of the nineteenth century. (In Caneva (1975) Grassmann's terminology in his physics publications is convincingly viewed as fitting in the manner of the Naturphilosophie although no evidence has been found that Grassmann saw himself in this tradition.) T. Hankins, in works in the general bibliography to this volume, has maintained the crucial nature of metaphysics in the creation of Hamilton's algebra. However, I essentially agree with David Bloor's evaluation at the Berlin Workshop on Social History of Mathematics, namely that, rather than being crucial, the metaphysical discussions by Hamilton were a dressing or 'glossing' of the mathematics and that more 'internal,' mathematical motivations were dominant.

Grassmann, in contrast, never implies that the metaphysical, or philosophical, was ever a part of the inspiration for the creation of the mathematics, not even for his extension of geometrical concepts to spaces of arbitrary (finite) dimension. But I hope it is evident on the basis of the summary description given above that Grassmann's philosophy is such an intimate part of his whole presentation in the _Ausdehnungslehre_ of 1844 that it is more than simply a philosophical interpretation of the mathematics, it should be treated also as a foundational program which relates the _Ausdehnungslehre_ to mathematics and science as a whole and provides a _wissenschaftlich_ justification for it. I think that historical analysis of this aspect of Grassmann's works is not so much a question of the extent of philosophical or theological influences on 'the mathematics' of Grassmann but rather of how Grassmann used his philosophical and theological knowledge and orientation in attempting to set up a new definition of mathematics itself. Perhaps Hamilton should be treated in the same way, but probably because his program is relatively so much less developed than Grassmann's, for example, it appears a more difficult task.

Bibliography

Caneva, K. (1975) _Conceptual and Generational Change in German Physics: The Case for Electricity, 1800-1846_. Princeton University, Ph. D. Dissertation.

Crowe, Michael J. (1967) _A History of Vector Analysis. The Evolution of the Idea of a Vectorial System_. Notre Dame (University of Notre Dame Press).

Grassmann, Hermann (1844) _Die Lineale Ausdehnungslehre, ein neuer Zweig der Mathematik, dargestellt und durch Anwendungen auf die übrigen Zweige der Mathematik, wie auch auf die Krystall-Mechanik, die Lehre vom Magnetismus und die Krystallonomie erläutert._ Leipzig.

Lewis, Albert C. (1975) _An Historical Analysis of Grassmann's Ausdehnungslehre of 1844._ University of Texas at Austin, Ph. D. Dissertation.

(1977) "H. Grassmann's 1844 _Ausdehnungslehre_ and Schleiermacher's _Dialektik_," _Annals of Science_ 34, 103-162.

APPENDIX

SOCIAL HISTORY OF MATHEMATICS

Herbert Mehrtens

1. "The sociology of mathematics concerns itself with the influence of forms of social organization on the origin and growth of mathematical conceptions and methods, and the rôle of mathematics as part of the social and economic structure of a period." This definition, taken from a classic paper by Dirk Struik (1942, 58), may well be used today for what is now called the social history of mathematics. To prepare ground for a thorough discussion of the field, however, a more detailed analysis is necessary. In the present paper, which is a strongly revised version of my contribution to the Berlin workshop, I shall attempt to outline the status, aims, object, methods, and problems of the social history of mathematics. References given in the text refer to the following "Select Bibliography". Although I have not attempted a critical survey of the literature, I hope this paper may also serve partially as an extended annotation to the bibliography.

Noting the frequent use of the words "sociology" and "social history" of mathematics as synonyms, it is necessary to clarify the difference right from the outset. For the sociologist historiographical work is empirical raw material used to test or exemplify abstract theoretical assertions. The historian's ultimate aim is to reconstruct an inter-connected series of events, attempting to render their historical unity understandable to the reader mainly by describing the actions and motives of individuals. Collectives, social structures and forces, and theoretical conceptions may well be involved. But in order to conserve the full complexity of the historical process and not to injure the individual dignity of the historical actor or event, the historian will always resort to some sort of common sense for his interpretation. Still there is a large meeting ground between sociology and history in their common aim of understanding the workings of the social processes in history. Nevertheless, neither historians nor sociologists are likely to yield to the professional standards and expectations of their counterparts. It is in these standards of the profession that the disciplines' aims find their expression. This basic difference in aims should be kept in mind when social history of mathematics is discussed.

A central feature of historiography is its narrative character. A series of events seen as determined by universal (sociological) laws is no longer a 'story' that can be told and could have happened otherwise. At this point a defensive move in the opposite direction is necessary, but can only be hinted at. The mathematician's understanding of the history of his subject is frequently sharply anti-sociological. But as frequently it is a 'rational reconstruction' of a development governed by universal laws - those of mathematics itself. In the unfolding of eternal mathematical laws through time there is no room for stories. They are only secondary flavouring, anecdotes of the lives of great men. The historiographer who sticks to the narrative and individualist nature of his discipline will have to guard his methodology against the rationales of mathematicians, sociologists and philosophers as well.

Turning to the present status of social history of mathematics, we find a valuable stock of studies in traditional historiography, biography, and the national and institutional history of mathematics. All three genera have their necessary place, and Biermann's history of mathematics at Berlin university (1973) may stand as one excellent example. In general, however, such studies are not very deep in analysis and rather one-sided in interpretation. The scope seems to broaden, however, at times when the social status of mathematics is being debated. This is, e.g., visible in the manifold activities within the educational reform movement at the turn of the twentieth century (for this movement cf. Inhetveen 1976, Pyenson 1979). Against the background of lively historiographical work in mathematics pieces of social history have been produced which well deserve the name and still are extremely valuable (e.g. Lorey 1916, C.Müller 19o4, Timerding 1914).

More recently a strong incentive for a social history of mathematics lay in the politically motivated, and mainly marxist, critique of the practice and ideology of mathematics.

Such attempts at critical historical analyses mainly remained fragmentary and stayed unpublished. But drawing on 'classic' marxist historiography of science like the works of Hessen, Struik, or Needham and using the developed historiographical tradition of Eastern marxist work (e.g. Wussing 1979), as well as the theoretical tools of a sophisticated and diversified Western marxism more intriguing results have turned up and may be expected (e.g. Hodgkin 1976).

In the professional historiography of science many strains of development, not only marxist analyses, have intertwined to a movement that has superseded the notorious internal-vs.-external debate. Social history of science, as depicted, say, in MacLeods report (1977) appears as a vast and colourful collection of studies that includes institutional and national histories as well as studies of specific scientific communities, small or large, where the connections between the knowledge produced and the social structure and development of the community are followed up. Studies concerning specific disciplines like mathematics are not in the majority but they form part of the picture. More general topics like those treated by Hahn (1971) or Forman (1971) present materials important for mathematics, while studies concerning the discipline directly present analyses of larger import (e.g. Folta 1977, Zetterberg 1977). In general, history of mathematics tends to produce more insight into social relations when historiography offers a closer analysis of the historical causes and roots of scientific developments. Hawkins' paper in this volume is an example of this point.

A further trend in social history of mathematics comes from neighbouring or 'meta'-disciplines. I have mentioned above sociology of science. Bloor's studies in this volume and elsewhere try to carry out the "strong programme" of sociology of science (1976, Ch.1) for mathematics, which could be taken as the attempt to show where in the realm between provable and necessary results and in individual choices of topics, methods, or concepts we can find social interests embedded in the style and content of mathematics. Fisher's studies of the "death" of invariant theory (1966, 1967) look at a theory as a social entity, to be analysed in terms of the attached group of mathematicians. Fisher's "death" has been repudiated because of the 'life' of invariant theory in the twentieth century[1]. This critique is due partly to Fisher's conceptual imprecision, partly to a misunderstanding of the term "theory" on the side of the critics. Mathematicians and historians may gain much for their historical understanding from such sociological studies, where the phenomena are differently conceptualized and new interpretations are presented.

A second field which contributes to social history of science is didactics, which seems to be turning to historical studies to a growing extent. The papers by Jahnke/Otte, Rogers

1) E.g. by Fang in his Sociology of Mathematics (Fang/Takayama 1975). This book, the only larger work bearing such a title, is of rather little help. It is a very preliminary attempt of the senior author to develop a sociological approach to mathematics, lacking most if not all the methodological sophistication of sociology, and of historiography as well.

and Schubring in this volume are examples of studies stemming from such a background. In regard to the fact that the main societal basis for mathematical work has been the educational system, the importance of studies from that point of view can hardly be underrated. A third field of influence is philosophy of science. As soon as an interest in the historical process of the production and dissemination of scientific knowledge developed within philosophy of science historical 'case studies' in this field opened new possibilities for historical analyses. Most important is certainly Lakatos' rational reconstruction of the communicative (and thus social) process of knowledge development (1973, 1976). Again we should be aware of the danger of misunderstanding, and of the fact that a 'rational' reconstruction is not sufficient as a historical reconstruction, especially so, when history, as with Lakatos, is banned to the footnotes.

The short survey of different types of studies and various disciplinary interests adding up to a stock of work in the social history of mathematics has shown this field to be more of a meeting place between disciplines than a homogeneous academic field. The interdisciplinary character of the Berlin workshop has proven the fertility of this ground. We have seen that the encounter between different methodological and theoretical approaches is able to add to the possibilities of historical understanding. On the other hand I have defended the unique approach of historical reconstruction against the dominance of 'rational reconstructions'. We find that the routes of theoretical science lead over our field and add to its fertility. But routes of traffic change in history. Can this meeting place remain a lively region, as part of the country called historiography, when other disciplines have stopped producing their 'case-studies' here and when theoretical tools are no longer brought to us but have to be imported? We should have a look at the nature of our grounds, the topography of the meeting place, the stock of tools, and at the connections with the larger area of history of science.

2. To find the place of social history within historiography of mathematics we have to talk about the object of the discipline and about the relative necessities, merits,and interconnections of different approaches to the historical analysis of this object. The object is 'mathematics', and since this phenomenon cannot be imagined without the mathematicians we find a simple necessity for the social-history-approach. The history of mathematicians is that of individuals, groups, institutions, and on a more sophisticated level of social roles, of processes of differentiation, autonomization, and professionalization. We further reach into matters of educational and science policy where our effort meets with that of general social history of science. There are, however, boundaries where the matter stops being simple.

A first problem lies in the term 'mathematician'. We tend to restrict the term to the person producing original mathematical knowledge. But this is inconsistent even with historiographical practice, when, e.g., the mathematical practitioners of the classic (Schneider 1970a, Taylor 1954, 1966) or the modern type (Gross, this vol.) or a larger mathematical culture is concerned (Pedersen 1963, Wallis 1972, 1980, Klemm 1958). Our implicit definition sorts out a certain type of knowledge, 'mathematics' in a wide sense, to determine groups of interest. The choice is, however, further determined by the relevance of the groups for knowledge production in mathematics. Practitioners and philomaths in the 18th century are signs of a wide social acceptance and dissemination of mathematics in relation to culture as well as to production.

The second boundary to the 'simple' social history, where definitions or restrictions in terms of knowledge enter the scene, lies in processes of basic social change. If we want to do more than state that a new social role for the mathematician was created in Germany during the early 19th century and describe university reform, the school system, the decline of academies, and the rise of seminars, we have to analyse those institutions and the role of mathematics therein in terms of the aims and restrictions imposed by the scholars themselves and by society at large. An institutional analysis, as exemplified in Schubring's work, will have to look at the knowledge produced and transmitted there. A thorough analysis cannot merely record actions and reactions of groups and individuals, it will have to refer to the knowledge connected with any science-related institution. On the other hand, the paper by Struik in this volume shows very clearly that any purely mathematical reconstruction of the great changes during the early 19th century will be insufficient. There may be legitimate pieces of purely social or purely disciplinary history. But sooner or later the connection between mathematics as a body of knowledge and mathematics as a field of social practice has to be taken into account.

We have to construe mathematics as both a body of knowledge and a field of social practice at the same time. These are not halves of a circular area embedded in the larger area equally divided into 'science' and 'society'.While the social practice of mathematics is determined by the nature of mathematics as a specific type of knowledge, the historical process of extension and change of mathematical knowledge is a social process inseperably embedded in the societal environment.An individual new idea in mathematics is brought forward as a 'knowledge claim'. This is an act of communication subject to specific social regulations. The evaluation of such a knowledge claim within the community of mathematicians again is a process of social interaction. It will depend the more on general ideas and norms concerning knowledge the more the innovation tends to leave well-worn paths. The inclusion of an innovation into the dogmatized body of taught mathematics, its dissemination into areas of application and other mathematical or scientific subdisciplines are social processes as well subject to regulations imposed by norms and institutions.

In such processes the structure of the acting collective (the mathematicians), of the institutions involved (journals, referees, meetings), of the stock of ideas at hand, of the ruling conceptions, aims, structures, and uses of mathematical knowledge are thoroughly intertwined. The latter elements, the normatively acting conceptions about knowledge, are one important strain connecting science and society. To institutionalize mathematical practice or new parts thereof in some place in society, the practitioners of the field have to prove their legitimacy. This will usually be implicit in an innovative process from the very beginning. The future tasks will be envisaged according to social standards concerning the man of knowledge. The complete set of the societal determination of the role of this type of knowledge and of its practitioners will play its part. The fact that the re-institutionalization of applied mathematics in German universities at the end of the 19th century was attempted through the introduction of applied mathematics in schools may serve to show how strongly innovations in mathematics of a larger institutional import in Germany had to prove legitimacy almost exclusively within the educational context (cf. Pyenson 1979).

I do not know of any useful definition of 'mathematics' in epistemological terms. Still, on those terms defenders of disciplinary rational reconstruction will point to the objectivity and coerciveness of mathematics and declare all analyses in terms of social phenomena to be secondary. This is a matter of philosophy of mathematics. For attempts to ground the sociological analysis on a specific philosophy of mathematics see Bloor (1973) and Phillips (1975). These authors argue that the usual 'realist' conception of mathematics is based on an implicit teleology and thus analytically unsound. Although I believe philosophy of mathematics to be necessary for historiography to some extent, I should rather argue that any philosophical prescription for historiography sets undue limits upon it. The historian has to be pragmatic about his conceptions of mathematics. Otherwise he will restrict his possibilities of approaching mathematics in different cultures. His understanding of mathematics taken from the present is the starting point. But in the hermeneutical approach this understanding will have to be modified according to the object of his study. It might become necessary to see mathematical reasoning as furnished with magical powers or to accept a different linguistic form of mathematics. Such traits need historical interpretation. The presentist interpretation, however, peeling out what is familiar mathematics and declaring the strange rest to be in need of explanation,is misleading. The underlying conception of mathematics is ahistorical, rendering all history into prehistory of the present.

History of mathematics is history of knowledge. But this history is a social process and 'knowledge' has to be taken in the widest sense of the term. Mathematics is a very specific knowledge. The implications of this specifity for historiography have to be a matter of further debate. The simple model of a hard core of objective knowledge and a soft belt of contemporary commitments, styles, and the like is certainly too simple. An open-minded historiography will help to widen our understanding of knowledge and mathematics, and of the social processes of their historical development.

3. In an attempt to provide a rough description of the intricate relations of knowledge and practice in mathematics I have made use of theoretical elements, e.g. in the concept of a 'knowledge claim' or in the discussion of the role of 'legitimacy' of a type of knowledge for institution building. Historiography is always guided by theoretical conceptions and history of mathematics even more so. The purely disciplinary reconstructions make use of modern mathematical theory in interpretation. Biography or annalistic history of institutions resorts to common-sense theories of human behaviour, and social history will have to take elements from the social science as theoretical tools. Any sophisticated historiography of science will have to make conscientious us of theories.

If we think, to take an example, about the development of mathematics, especially of geometry, in Germany between 1800 and 1870 we can proceed along different routes laid out by theoretical orientations. The first and natural approach would be through disciplinary mathematical reconstruction. We can describe the emergence and mathematical roots of various new approaches and methods in geometry. We will note their interaction and the influence of other mathematical developments and thus may neatly order and connect a sequence of mathematical ideas. But at certain points this will remain insufficient. We may explain the receptivity for non-Euclidian geometry by the interaction of different approaches, specifically noting the role of differential geometry. But how to account for the emergence of the idea that a non-Euclidean geometry is legitimate? And, as an important parallel, how to interpret Grassmann's new conception of geometry in his Ausdehnungslehre of 1844? The latter question has been thoroughly analyzed in its relation to Schleiermacher's philosophy by Lewis. For changing concepts of space and of the relation of geometry to natural space we might well look for philosophical influences. Here is another general approach of a more idealist nature: to see mathematics as a mainly autonomous intellectual endeavour, which is at certain important stages influenced or shaped by more general scientific or philosophical ideas. This will still not suffice to answer all questions. How about the roots of descriptive and projective geometry? At this point a third approach might start. Classical marxist analysis (e.g. Hessen 1931) relates the history of science to productive forces and modes of production. The emergence of descriptive geometry may be construed as the codification of basically elementary mathematical knowledge in reaction to the needs of production under certain institutional imperatives (the Ecole Polytechnique, cf. Paul 1980). Projective geometry then emerged as the mathematical analysis of that code of knowledge. Similarly the role of geodesy in those times and the engagement of mathematicians (e.g. Gauss) in such practical problems may be seen as another causal factor to mathematical developments. Still another perspective is opened up when we view mathematicians as thinking about their subject under changed institutional and social conditions Making use of theories of social change and especially of systems theory in functionalist sociology one might focus on the process of disciplinary differentiation and autonomization in that period (e.g. Stichweh 1977) and mark the growing distance between mathematics and physics, the necessity of more disciplinary self-consciousness in periods of marked change as in the early 19th century, the new situation of mathematics as fully integrated in the educational system, and the ensuing pressure to rethink basic concepts and methods in terms of the educational task in a system of not profession-oriented elite education. Still other approaches may speculate about the 'social imagery' (Bloor 1976) inherent in the new geometry in a period of the shaping of bourgeois society, try to embed mathematics into the general change of world-views, attempt a milieu-theoretical analysis, or may even analyse the new type of rationality emerging in that period (cf. the rather abstract discussion in Jahnke/Otte 1981, Introduction).

It should be clear that any of the approaches speculated upon has its own right and may produce an important part of the picture. But any single-line reconstruction will leave considerable gaps in analysis and understanding. Thus we see the need for theories opening new vistas and explaining developments as well as the impossibility of placing all the historical material in the shelves of one theoretical cabinet. If we want to end up with a historical narrative which is meaningful and convincing to fellow historians and to a more general public including mathematicians the 'meaning' of the story and its elements cannot be drawn from one disciplinary background. Thus the best advice to the historian might be to ruthlessly exploit the offerings of theoretical disciplines while scruplously checking the applicability and explanatory range of the pieces used against his empirical material. This should also be a plea for the value of single-lined and even speculative approaches - as long as the limitations are clear to author and reader. Historiography itself is a social process, and specific studies will add up to shape and criticise, and reshape again the general picture.

4. The use of theories to identify important phenomena, to rate their respective influence on the course of events, and to construe the historical connex of events is a method for the historian to achieve his aim. Methods in a stricter sense, prescriptions for answering specifical historical questions are constitutive to historiography. The brief discussion of general approaches suggests that to develop a systematic methodology for social history of mathematics will be a very difficult task. I shall only be able to give some hints in this direction.

A sociologically oriented historiography will lay stress on collective actors, on the analysis of institutions, on a critique of ideology, and on the embedding of specific events in general social development. Historical analysis of an event will give a description of the event, analyse its causes and effects, and evaluate the conditions for the possibility of the event. As to historical actors we will look at individuals, collectives, institutions, and nations. Finally, looking at knowledge, we can analyse its history into invention, innovation, dissemination, and application. Although this list of categories is incomplete, and its members are not independent of one another, it could nonetheless be used as a conceptual grid to classify types of historical problems and discuss adequate methods of solving them. For lack of space and for systematic reasons I shall not attempt to present a methodological inventory here. Instead I shall try to exemplify the procedures of social history of mathematics in more detail, discussing social entities and phases of the development of knowledge in combination. Many important approaches, methods, or types of analysis will have to be left out. It should be added that the concepts applied are not stiff frames for historical pictures. They are meant to be exploited in their floating meaning and their inherent variability.

At the core of classical historiography of mathematics lies the study of individual achievements, mainly inventions of knowledge. The social counterpart is given by biographical studies of different kinds. The term 'invention' is used to describe the individual production of a knowledge claim contrasting it analytically with the process of 'innovation', i.e. the introduction of a new piece into the body of knowledge accepted by some larger collective. The distinction between invention and innovation rests on a conception of the process of mathematical knowledge development as an interaction of the individual who creates new knowledge and the collective which rejects, accepts, or modifies the invention in whole or in part and thus transforms the body of commonly accepted knowledge. While the individual is bound to the knowledge, the norms and the forms of interaction of the collective, the dynamics of the collective rests on individual actions and individual deviation. This dialectical nature of the individual-collective interaction requires a complementary analysis of knowledge development in terms of both social and epistemological processes. In terms of method this leads to the procedure of analysing an individual invention by segregating those parts which may be explained on the basis of the contemporary knowledge of the scientific community as 'normal' steps, taking common standards, aims, and methods into account, from the 'extraordinary' parts (cf. Mehrtens 1978, 200). These latter elements of an invention have then to be explained by extraordinary elements of the social structure of the scientific community, by the influence of other collectives with which the individual interacts, by singular 'influences', and/or individual biography. Putting the individual into contexts other than that of his scientific community makes necessary a partly sociological analysis or at least - since the _explanans_ is rated as 'extraordinary' - mention of the social conditions that have made possible such factors in intellectual development. Thus the analysis of individual achievements leads into a contextual biographical study (e.g. Bloor, this vol., Lewis 1975, Norton 1978). Biographical studies as such supply materials for such an analysis of achievements. In terms of social history, furthermore, they provide knowledge of the material basis of collective and individual life describing family background and education, social role and social status of the mathematician, showing his place in the contemporary intellectual network,and the like. Of special importance for the general social history of mathematics are studies of those individuals who have been effective in the social development of mathematics. Eccarius' studies on Crelle may stand as an excellent example. Studies concerning the organizational activities of Felix Klein at the end of the century may be taken as typical as well, although they merely incorporate biographical elements (Manegold 1970, Pyenson 1979, Tobies 1979).

The study of collectives is even more complicated for the attempt at an abstract methodological description. Collectives clearly can be found and have to be studied on many different levels. The most important collectives for our endeavour are those which may be seen as historical actors in the history of mathematical knowledge: groups with a special communicative connection, disciplinary communities, and the mathematical community at large. As for the individual a kind of contextual biography of the group is necessary. The methodological approach would be 'prosopography' (cf. Pyenson 1977), i.e. the collective biography of the group, noting elements of social and intellectual homogeneity and diversity, analysing the historical context and the intellectual nature of group commitments, studying social structure, societal locus, and the social and intellectual interests of the group. Studies of groups encorporating such elements of analysis may be found in the work of Enros (1979), Folta (1977), Mehrtens (1981), and others.

A second kind of collective may be called the social 'neighbourhoods' of individuals or actor-collectives. These are of special importance in the analysis of mathematical innovation. While we may see the process of reception of an invention by the disciplinary community concerned as a 'normal' process governed by the contemporary body of knowledge and commitments, we frequently notice extraordinary processes of considerable transformation, disregard or rejection of an invention. Again the extraordinary needs special explanation, which regularly has to resort to the larger context and look at specific 'neighbourhoods'. Bloor's contextualization of Hamilton (this vol.), Richards' study of the reception of non-Euclidean geometry in England (1979), or MacKenzie's analysis of a controversy in statistics (1978) may serve as examples for such an analysis. An 'archeological' analysis of contemporary language and forms of conceptualization (as Hodgkin suggests, this vol.) or a synchronic study of different scientific disciplines or even cultural regions might be adequate. This could be done by following the track of certain concepts. The 'organic' synthesis, as attempted by Steiner's geometry, may, e.g., be related to the role of the concept of'organism' in the intellectual ambience of early 19th century Germany. Or the methodological prominence of 'conceptual structuring' in disciplines as diverse as mathematics and the legal sciences in the mid 19th century may be exploited to see and explain mathematics as part of a larger intellectual neighbourhood.

Group-commitments play a strong role in an internal analysis of knowledge development where 'schools' come in. Hawkins (1980, this vol.) gives a fine example of how an analysis of the historical factors involved in an individual's mathematical achievement leads to recognizing the importance of 'school'-connections.

When speaking of 'institutions' I shall keep to the stricter sense of the term, meaning fixed and visible social structures like the Königsberg seminar, the Prussian university Crelle's journal, or a scientific society. As existing, basically unchanged institutions they provide a social locus for mathematical activities, present boundary conditions for innovation, dissemination and application of knowledge, and encorporate and express societal conditions for mathematical practice. As such they should be analysed. Here elements of the sociology of institutions will provide important methodological aids. Of high importance for the social history of mathematics proper and for the social process of knowledge development are the creation, modification, and abolition of institutions. The change of university mathematics in Germany from utilitarian instruction to the training of research-oriented mathematicians is such a development, one of high importance (cf. Dauben 1981, Langhammer 1981, Lorey 1916, Turner 1971). A further approach lies in various sorts of functional analyses. The Königsberg seminar of Jacobi and

Neumann, e.g., may be viewed both in terms of its role in the professionalization and autonomization of mathematics, and in terms of its educational task, divided again into the transmission of specific knowledge and its role as part of the general system of neo-humanist elite education. In such an analysis of institutions lies one important approach to the problem of the mediation of social needs of practical or ideological nature into the work of a specific discipline.

Nations or similar extensive (and complex) social units shall finally be remarked upon. Nations come into view when the historical change of scientific world-centers (as from France to Germany in the 19th century) is analysed, or when mathematics is embedded in the larger neighbourhood of a scientific or intellectual culture. Bottazzini's paper (this vol.) may be taken as an example of how national and mathematical developments interact in certain periods. Similarly a general history of mathematics in Germany in the 19th century will have to take into account the attempts for national unification, the leading political and intellectual role of Prussia, the role and connex of Bildung (education) and Besitz (property) in Germany's history. As a description of a very general development, such a history will have to make use of general histories of all kinds as well as of histories of movements, institutions, and ideas that play a role in mediating national specifics into the history of mathematics. An even more important methodological approach, though rarely carried out, is the comparative analysis of national developments (for an example on a very general level cf. Needham 1956). This would be of special importance for the history of early 19th century mathematics. Many papers in this volume and in the collection edited by Jahnke and Otte (1981) make clear how strongly this was a period of social and intellectual transformation against the background of the industrial and political revolutions. Educational and epistemological philosophies played a strong but varying role in these developments in different countries. Comparative analyses would help much to clear up causes and context of disciplinary change in this and similar cases (for a brief comparison of France and Prussia cf. Schubring 198o). As a last mode of analysis on a very general level, which might be relevant to disciplinary history, we might mention Wuthnow's attempt to analyse the political and economical world system in the 17th and 18th centuries as favourable to the development of science (1979).

Returning to the question of the structure and place of social history of mathematics, I would like to stress the aim of viewing mathematics 'in context'. It provides the most important more general 'neighbourhood'. While social history of mathematics proper, i.e. that part of historiography of mathematics that focusses on social entities and views mathematics mainly as a social activity, has its necessary place within the larger disciplinary context, its present role for the development of historiography lies in the fact that its advocates are pressing for contextualisation as a means of historical understanding. Hence it is closely related to all attempts, including those of purely intellectual history, which tend to leave behind the traditional combination of a disciplinary, rational reconstruction of history in terms of mathematics with a historical record of the history of mathematicians of little interpretational import.

To end this paper, I shall briefly summarize what I find to be the most general problems for the social history of mathematics. I have just hinted at the problem of demarcation of the field. We can, by describing theory, method, aim,and object, try to give a definition. This will bring out the characteristics of what I call the social history of mathematics proper. But, as we have seen, the borderland is vast; its limits itself are floating. It seems much more important at the moment to keep alive and fruitful the meetingplace of different historiographical and theoretical interests than to build a well defined subdiscipline. A second problem has been stated during the discussions of the workshop, namely the relation of the historian to his public. The strange lack of studies in the interaction of mathematics and physics seems to result from the traditional relation to the public. Historians of physics mainly adress physicists and for historians of mathematics the mathematicians are the adressee. As professionalized historiography of science is gaining a certain independence from the specific disciplinary environment social history of mathematics will even more fail to fulfil the expectations of mathematicians. As it tends to encorporate critique of ideologies as part of historical analysis it will conflict with present ideologies of mathematics because it then ceases in part to fulfil the usual and traditional legitimating function of history. This is necessary to achieve a better historical understanding, and it is, I believe, necessary so that historiography of science may find its place in the task of producing a human future. But we have to know that we are in a situation of conflicting interests and not only of competing knowledge claims.

The last two problems are of a more epistemological nature. We do not have adequate conceptions to analyse human activity in the history of knowledge in its double nature as social and intellectual process. Any study in the social history of mathematics adressing that relation is a piece of practice to a theory yet to be developed. As part of the social process of attaining knowledge, it ought to be done with conscious reflexiveness. The modes of analysis applied by the historian ought to be applicable and be applied to his own professional practice in order to allow a rational discourse on history and historiography. This is certainly difficult. And the last problems adds to this difficulty. Our theoretical understanding of the relation of science and society is quite imperfect. In this respect as well we have to note that we are not working on a safe basis of corroborated theoretical understanding but with a mixture of theoretical elements and often questionable common-sense conceptions. We have to see our work as part of an endeavour aimed at gaining such an understanding, which,at the same time, is an understanding of our own work in the context of our world and society.

SELECT BIBLIOGRAPHY

Compiled by Herbert Mehrtens

The bibliography attempts a survey of works on the social history of mathematics, mainly, but not exclusively, of the 19th century. Criteria for the selection of entries must remain as vague as the boundaries of the field. I have limited myself, with few exceptions, to mathematics proper, and have been restrictive with biographies and institutional histories. General works on the history of mathematical knowledge have been included in a few cases of special relevance to social history. I gratefully acknowledge the help of many colleagues. The responsibility, however, for omissions and doubtful entries is entirely mine. There certainly remains a personal bias in the selection, e.g. towards studies of German provenance.

Abbreviations

HM	Historia Mathematica
IDM Mat.	Institut für Didaktik der Mathematik der Universität Bielefeld: Materialien und Studien
NTM	NTM Schriftenreihe für Geschichte der Naturwissenschaften, Technik und Medizin
ZDM	Zentralblatt für Didaktik der Mathematik

Further abbreviations follow _Isis Critical Bibliography_.

Andersen, Kirsti
1980 "An impression of Mathematics in Denmark in the Period 1600 - 1800." Centaurus 24(1980), 316-334.

Ball, Walter W.R.
1889 A History of the Study of Mathematics at Cambridge. Cambridge (Cambridge Univ. Pr.).

Bekemeier, Bernd
1980 "Zum Zusammenhang von Wissenschaft und Bildung am Beispiel des Mathematikers und Lehrbuchautors Martin Ohm." In: Høyrup/Bekemeyer (1980), 139-234.

Ben-David, Joseph
1971 The Scientist's Role in Society. A Comparative Study. Englewood Cliffs (Prentice-Hall).

Bernhardt, Hannelore
1980 "Zur Institutionalisierung der angewandten Mathematik an der Berliner Universität 1920 - 1933." NTM 17(1)(1980), 23-31.

Biermann, Kurt-R.
1959a Johann Peter Gustav Lejeune Dirichlet - Dokumente für sein Leben und Wirken.(Abh. Deut.Akad.Wiss., Kl.Math.Phys. Tech. 1959, Nr.2) Berlin (Akademie Verlag).
1959b "Über die Förderung deutscher Mathematiker durch Alexander von Humboldt." In: Alexander von Humboldt Gedenkschrift. Berlin (Akademie Verlag), 83-159.
1960 Vorschläge zur Wahl von Mathematikern in die Berliner Akademie. (Abh. Deut.Akad.Wiss.,Kl.Math. Phys.Tech. 1960, Nr.3) Berlin (Akademie Verlag).
1973 Die Mathematik und ihre Dozenten an der Berliner Universität 1810-1920. Berlin (Akademie Verlag).

Bloor, David
1973 "Wittgenstein and Mannheim on the Sociology of Mathematics." Stud.Hist.Phil.Sci.4(1973), 173-191.
1976 Knowledge and Social Imagery. London (Routledge and Kegan Paul).
1978 "Polyhedra and the Abominations of Leviticus." Brit.J.Hist.Sci. 11(1978), 245-272.

Bockstaele, P.P.

1978 "Mathematics in the Netherlands from 1750 to 1830." _Janus_ 65(1978), 67-95.

Bos, Henk

1977 "Calculus in the Eighteenth Century - The Rôle of Applications." _Bull.Inst.Math.Appl._ 13(1977), 221-227.

1978 "Was lehren uns historische Beispiele über Mathematik und Gesellschaft." _ZDM_ 10(1978), 69-75.

Bos, Henk and Herbert Mehrtens

1977 "The Interactions of Mathematics and Society in History. Some Exploratory Remarks." _HM_ 4(1977),7-30.

Bottazzini, Umberto

1977 "Riemanns Einfluß auf E.Betti und F.Casorati." _Arch.Hist.Exact Sci._ 18(1977), 27-37.

Brentjes, Sonja

1977 _Untersuchungen zur Geschichte der linearen Optimierung (LO) von ihren Anfängen bis zur Konstituierung als selbständige mathematische Theorie._ Diss. Univ. Dresden.

Brock, William H.

1975 "Geometry and Universities: Euclid and His Modern Rivals 1860 - 1901." _Hist.Educ._ 4(2)(1975), 21-36.

Brock, William H. and Roy MacLeod

1980 _The Journals of Thomas Archer Hirst FRS_. London (Mansell).

Chapelon, Jacques

1971 "Mathematics and Social Change." In: Lelionnais, F. (Ed.), _Great Currents of Mathematical Thought._ 2 vols., New York, 212-221.

Christmann, Erwin

1925 _Studien zur Geschichte der Mathematik und des mathematischen Unterrichts in Heidelberg._ Diss. Univ. Heidelberg.

Condorcet, Marie-Jean-A.-N. de

1974 _Mathématique et société_. Choix des textes et commentaire par R. Rashed. Paris.

Cowan, Ruth S.

1972 "Francis Galton's Statistical Ideas: The Influence of Eugenics." Isis 63(1972), 509-528.

1977 "Nature and Nurture: Biology and Politics in the Work of Francis Galton." Stud.Hist.Biol. 1(1977), 133-208.

Crane, Diana

1972 Invisible Colleges. Diffusion of Knowledge in Scientific Communities. Chicago (Univ. of Chicago Pr.

Crosland, Maurice P.

1967 The Society of Arcueil. A View of French Science at the Time of Napoleon. London (Heinemann).

1975 (Ed.) The Emergence of Science in Western Europe. London (MacMillan).

Daston, Lorraine J.

1979 The Reasonable Calculus: Classical Probability Theory, 1650 - 1840. Diss. Harvard Univ.,Cambridge.

1980 "Probabilistic Expectation and Rationality in Classical Probability Theory." HM 7(1980), 234-260.

1981 "Mathematics and the Moral Sciences: The Rise and Fall of the Probability of Judgements, 1785 - 1840." In: Jahnke/Otte (1981), 287-309.

Dauben, Joseph W.

1977 "Georg Cantor and Pope Leo XIII: Mathematics, Theology and the Infinite." J.Hist.Ideas 38(1977), 85-108.

1980 "Mathematicians and World War I: The International Diplomacy of G.H.Hardy and Gösta Mittag-Leffler as Reflected in Their Personal Correspondence." HM 7(1980), 261-288.

1981 "Mathematics in Germany and France in the Early 19th Century: Transmission and Transformation." In: Jahnke/Otte(1981), 371-399.

Dubbey, John M.

1978 The Mathematical Work of Charles Babbage. Cambridge (Cambridge Univ. Pr.).

Dugac, Pierre

1976 Richard Dedekind et les fondements des mathématiques. Paris (Vrin).

Eccarius, Wolfgang

1974 Der Techniker und Mathematiker August Leopold Crelle (1780-1855) und sein Beitrag zur Förderung und Entwicklung der Mathematik im Deutschland des 19. Jahrhunderts. Diss. Leipzig. Abstract (same title) in NTM 12(2)(1975), 38-49.

1976a "August Leopold Crelle als Herausgeber wissenschaftlicher Fachzeitschriften." Ann.Sci. 33(1976), 229-261.

1976b "August Leopold Crelle als Herausgeber des Crelleschen Journals." J.reine angew.Math. 286/287(1976), 5-25.

1977a "Zur Gründungsgeschichte des Journals für die reine und angewandte Mathematik." NTM 14(2)(1977), 1-7.

1977b "August Leopold Crelle als Förderer bedeutender Mathematiker." Jahresber.Deut.Math.Ver. 79(1977), 137-174.

1977c "August Leopold Crelle und der Versuch einer Reorganisation des preußischen Mathematikunterrichts in der Periode der industriellen Revolution." Mitteilungen der math. Gesellschaft der DDR 1977, Heft 1, 90-95.

1980a "Der Gegensatz zwischen Julius Plücker und Jakob Steiner im Lichte ihrer Beziehungen zu August Leopold Crelle." Ann.Sci. 37(1980), 189-213.

1980b "Die wissenschaftliche Gemeinschaft der Mathematiker im Deutschland des 19.Jahrhunderts.(Bemerkungen zu einer wünschenswerten deskriptiven Statistik)." Rostocker wissenschaftshistorische Manuskripte Heft 5(1980), 59-68.

Eichberg, Henning

1974 "'Auf Zoll und Quintlein' Sport und Quantifizierungsprozess in der frühen Neuzeit." Arch.Kulturgesch. 56 (1974), 141-176.

1977 "Geometrie als barocke Verhaltensnorm. Fortifikation und Exerzitien." Z.hist.Forsch. 4(1977), 17-50.

Enros, Philip C.

1979 The Analytical Society: Mathematics at Cambridge University in the Early Nineteenth Century. Diss. Univ. of Toronto.

Fang, J. and K.P. Takayama

1975 Sociology of Mathematics and Mathematicians - A Prolegomenon. Happauge, N.Y. (Paideia Pr.).

Fisher, Charles S.

1966 "The Death of a Mathematical Theory: A Study in the Sociology of Knowledge." Arch.Hist.Exact Sci. 3 (1966/67), 137-159.

1967 "The Last Invariant Theorists." Archives Européennes de sociologie 8(1967), 216-244.

1972 "Some Social Characteristics of Mathematicians and Their Work." Amer.J.Sociol. 78(1972/73), 1094-1118.

Folkerts, Menso

1974 "Die Entwicklung und Bedeutung der Visierkunst als Beispiel der praktischen Mathematik der frühen Neuzeit." Hum.Tech. 18(1974), 1-41.

Folta, Jaroslav

1977 "Social Conditions and the Founding of Scientific Schools. An Attempt at an Analysis on the Example of the Czech Geometric School." Acta hist.rerum natur.tech. , Special issue 10 (1977), 81-179.

Folta, Jaroslav and Lubos Nový

1965 "Sur la question des méthodes quantitatives dans l'histoire des mathématiques." Acta hist.rerum nat.tech., Special issue 1 (1965), 1-35.

Forman, Paul

1971 "Weimar Culture, Causality, and Quantum Theory, 1918 - 1927. Adaption by German Physicists and Mathematicians to a Hostile Intellectual Environment." Hist.Stud.Phys.Sci. 3(1971), 1-115.

Fry, Thornton C.

1964 "Mathematicians in Industry - The First 75 Years." Science 143(1964), 934-938.

(Gauss)

1955 C.F.Gauss und die Landesvermessung in Niedersachsen. Hannover.

Gericke, Helmuth

1955 Zur Geschichte der Mathematik an der Universität Freiburg i.Br. (Beiträge zur Freiburger Wissenschafts- und Universitätsgeschichte 7) Freiburg.

Gerstell, Marguerite

1975 "Prussian Education and Mathematics." Amer.Math. Mon. 82(1975), 240-245.

Gillispie, Charles C.

1972 "Probability and Politics: Laplace, Condorcet, and Turgot." Proc.Amer.Phil.Soc. 116(1972), 1-20.

Grabiner, Judith V.

1974 "Is Mathematical Truth Time-Dependent?" Amer.Math. Mon. 81(1974), 354-365.

1981 "Changing Attitudes Toward Mathematical Rigor: Lagrange and Analysis in the Eighteenth and Nineteenth Centuries." In: Jahnke/Otte (1981), 311-347.

Grattan-Guinness, Ivor

1981a "Mathematical Physics in France: Knowledge, Activity and Historiography." In: Dauben, J.W. (Ed.), Mathematical Perspectives: Essays on Mathematics and its Historical Development. New York (Academic Pr.) (in print).

1981b "Mathematical Physics in France, 1800 - 1835." In: Jahnke/Otte (1981), 349-370.

Gross, Horst-Eckart

1978 "Das sich wandelnde Verhältnis von Mathematik und Produktion." In: Plath,P. and H.J.Sandkühler (Eds.), Theorie und Labor. Dialektik als Programm der Naturwissenschaft. Köln (Pahl-Rugenstein), 226-269.

Grundel, Friedrich

1928 Die Mathematik an den deutschen höheren Schulen. 2 vols., Leipzig (Teubner) 1928/29.

Hahn, Roger

1971 The Anatomy of a Scientific Institution. The Paris Academy of Science, 1666 - 1803. Berkeley (Univ. of California Pr.).

1975 "New Directions in the Social History of Science." Physis 17(1975), 205-218.

Hankins, Thomas L.

1977 "Triplets and Triads: Sir William Rowan Hamilton on the Metaphysics of Mathematics." Isis 68(1977), 175-193.

Harig, Georg

1958 "Über die Entstehung der klassischen Naturwissenschaften in Europa." Deut.Z.Philos. 6(1958), 519-550.

Hawkins, Thomas

1977 "Weierstrass and the Theory of Matrices." Arch.Hist. Exact Sci. 17(1977), 119-163.

1980 "Non-Euclidean Geometry and Weierstrassian Mathematics: The Background to Killing's Work on Lie Algebras." HM 7(1980), 289-342.

Hessen, Boris

1931 "The Social and Economic Roots of Newton's Principia." In: Bukharin, N.I. et al., Science at the Cross Roads. London, new ed. London (F.Cass) 1971, 149-212.

Hodgkin, Luke

1976 "Politics and Physical Sciences." Rad.Sci.J. 4(1976), 29-60.

Hofmann, Joseph E.

1959 "Alexander von Humboldt in seiner Stellung zur reinen Mathematik in ihrer Geschichte." In: Alexander von Humboldt Gedenkschrift. Berlin (Akademie Verlag), 237-287.

Høyrup, Else

1978 Women and Mathematics, Science, and Engineering. A Bibliography. Roskilde (Univ. Library).

1979 Books About Mathematics. History, Philosophy, Education, Models, Systems Theory, and Works of Reference etc. A Bibliography. Roskilde (Univ. Center).

Høyrup, Else and Jens Høyrup
1973 Matematikken i samfundet. Copenhagen (Gyldendals Samfundsbibliothek).

Høyrup, Jens
1980 "Influences of Institutionalized Mathematics Teaching on the Development and Organization of Mathematical Thought in the Pre-Modern Period." In: Høyrup/Bekemeyer (1980), 7-137.

Høyrup, Jens and Bernd Bekemeier
1980 Studien zum Zusammenhang von Wissenschaft und Bildung. (IDM Mat. 20) Bielefeld.

Inhetveen, Heide
1977 "Zur Geschichte des Mathematikunterrichts." In: Rieß, F. (Ed.), Kritik des mathematisch-naturwissenschaftlichen Unterrichts. Frankfurt (Päd. extra), 139-206.

1976 Die Reform des gymnasialen Mathematikunterrichts zwischen 1890 und 1914 - Eine sozioökonomische Analyse. Bad Heilbrunn (Klinkhardt).

Jahnke, Hans Niels and Michael Otte (Eds.)
1981 Epistemological and Social Problems of the Sciences in the Early Nineteenth Century. Dordrecht (Reidel).

Jungnickel, Christa
1979 "Teaching and Research in the Physical Sciences and Mathematics in Saxony, 1820 - 1850." Hist.Stud.Phys.Sci. 10(1979), 3-47.

Keller, Alex
1972 "Mathematical Technologies and the Growth of the Idea of Technical Progress in the Sixteenth Century." In: Debus, A.(Ed.), Science, Medicine and Society in the Renaissance. New York, 11-27.

1975 "Mathematicians, Mechanics, and Experimental Machines in Northern Italy in the Sixteenth Century." In: Crosland (1975), 15-34.

Klemm, Friedrich
1958 "Die Berliner Philomathische Gesellschaft (Philomathie)." Sudhoffs Arch. 42(1958), 39-45.

Klemm, Friedrich

1966 "Die Rolle der Mathematik in der Technik des 19. Jahrhunderts." Technikgesch. 33(1966), 72-90.

Koppelmann, Elaine

1971 "The Calculus of Operations and the Rise of Abstract Algebra." Arch.Hist.Exact Sci. 8(1971), 155-242.

Kuzawa, Mary Grace

1966 The Origin and Development of the Polish School of Mathematics (1918-1939). Diss.New York Univ.

Laita, Lewis M.

1975 A Study of the Genesis of Boolean Logic. Diss. Notre Dame Univ.

1977 "The Influence of Boole's Search for a Universal Method of Analysis on the Creation of his Logic." Ann.Sci. 34(1977), 163-176.

Lakatos, Imre

1973 "History of Science and its Rational Reconstruction." Boston Stud.Phil.Sci. 8(1973), 91-136.

1976 Proofs and Refutations. The Logic of Mathematical Discovery. Cambridge (Cambridge Univ.Pr.).

Lewis, Albert C.

1975 An Historical Analysis of Grassmann's Ausdehungslehre of 1844. Diss.Univ.of Texas, Austin.

1977 "H.Grassmann's 1844 Ausdehnungslehre and Schleiermacher's Dialektik." Ann.Sci. 34(1977), 103-162.

1981 "Justus Grassmann's School Programs as Mathematical Antecedents of Hermann Grassmann's 1844 'Ausdehnungslehre'." In: Jahnke/Otte (1981), 255-267.

Langhammer, Walter

1981 "Some Aspects of the Development of Mathematics at the University of Halle-Wittenberg in the Early 19th Century." In: Jahnke/Otte (1981), 235-254.

Lighthill, J.

1976 "The Interaction Between Mathematics and Society." Bull.Inst.Math.Appl. 12(1976), 288-299.

Lindner, Helmut

1980 "'Deutsche' und 'gegentypische' Mathematik. Zur Begründung einer 'arteigenen' Mathematik im 'Dritten Reich' durch Ludwig Bieberbach." In: Mehrtens,H. and S. Richter (Eds.), Naturwissenschaft, Technik und NS-Ideologie. Frankfurt (Suhrkamp),88-116.

Lorey, Wilhelm

1916 Das Studium der Mathematik an den deutschen Universitäten seit Anfang des 19.Jahrhunderts. Leipzig (Teubner).

Mackensen, Ludolf von

1969 "Bedingungen für den technischen Fortschritt, dargestellt anhand der Entwicklung und ersten Verwertung der Rechenmaschinenerfindung im 19.Jahrhundert." Technikgeschichte 36(1969), 89-102.

MacKenzie, Donald

1978 "Statistical Theory and Social Interests: A Case-Study." Soc.Stud.Sci. 8(1978), 35-83.

1979 "Karl Pearson and the Professional Middle Class." Ann.Sci. 36(1979), 125-143.

1981 Statistics in Britain 1865-1930. The Social Construction of Knowledge. Edinburgh (Edinburgh Univ. Press).

MacLeod, Roy

1977 "Changing Perspectives in the Social History of Science." In: Spiegel-Rösing, I. and D. de Solla-Price (Eds.), Science, Technology and Society: A Cross-Disciplinary Perspective. London (Sage), 149-195.

Manegold, Karl Heinz

1966 "Eine Ecole Polytechnique in Berlin." Technikgesch. 33(1966), 182-196.

1968 "Felix Klein als Wissenschaftsorganisator." Technikgesch. 35(1968), 172-204.

1970 Universität, Technische Hochschule und Industrie. Ein Beitrag zur Emanzipation der Technik im 19. Jahrhundert unter besonderer Berücksichtigung der Bestrebungen Felix Kleins. Berlin (Duncker&Humblot).

May, Kenneth O.

1968 "Growth and Quality of Mathematical Literature." Isis 59(1968), 363-371.

1973 Bibliography and Research Manual of the History of Mathematics. Toronto (Univ. of Toronto Pr.).

Mehrtens, Herbert

1976 "T.S.Kuhn's Theories and Mathematics: A Discussion Paper on the 'New Historiography' of Mathematics." HM 3(1976), 297-320.

1977 "Drei Beispiele zur Sozialgeschichte der Mathematik." Materialien zur Analyse der Berufspraxis des Mathematikers (Bielefeld) 19(1977), 129-138.

1978 "Bemerkungen zur pragmatischen Philosophie, Sozial- und Ideengeschichte der Mathematik am Beispiel der Entstehung der Verbandstheorie." In: Steiner, H.G. (Ed.), Zum Verhältnis von Mathematik und Philosophie im Unterricht der Sekundarstufe II/Kollegstufe. (IDM Mat. 12) Bielefeld, 189-210.

1981 "Mathematicians in Germany circa 1800." In: Jahnke/ Otte (1981), 401-420.

Merton, Robert K.

1938 "Science, Technology and Society in 17th Century England." Osiris 4(1938), 360-632; repr. with new introduction by the author, New York 1970.

Müller, Conrad H.

1904 Studien zur Geschichte der Mathematik, insbesondere des mathematischen Unterrichts an der Universität Göttingen im 18. Jahrhundert. Diss. Univ. Göttingen.

Neander, Joachim

1974 Mathematik und Ideologie. Zur politischen Ökonomie des Mathematikunterrichts. Starnberg (Raith).

Needham, Joseph

1956 "Mathematics and Science in China and the West." Sci.Soc. 20(1956), 320-343.

Nielsen, Niels

1929 Géomètres Francais sous la revolution. Copenhagen (Levin&Munksgaard).

Norton, Bernard
1978 "Karl Pearson and Statistics: The Social Origins of Scientific Innovation." Soc.Stud.Sci. 8(1978),3-34.

O'Hara, James G.
1979 Humphrey Lloyd (1800-1881) and the Dublin Mathematical School of the Nineteenth Century. Diss. Univ. of Manchester.

Osen, Lynn M.
1974 Women in Mathematics. Cambridge, Mass.(MIT Pr.).

Otte, Michael
1977 "Zum Verhältnis von Wissenschafts- und Bildungsprozeß - dargestellt am Beispiel der Entwicklung der Mathematik im 19.Jahrhundert (Beschreibung eines Projekts)." ZDM 9(1977), 205-209.

Pahl, Franz
1913 Geschichte des naturwissenschaftlichen und mathematischen Unterrichts. Leipzig (Quelle&Meyer).

Paul, Matthias
1980 Gaspard Monges 'Géométrie Déscriptive' und die Ecole Polytechnique - Eine Fallstudie über den Zusammenhang von Wissenschafts- und Bildungsprozess. (IDM Mat. 17) Bielefeld.

Pearson, Karl
1978 The History of Statistics in the 17th and 18th Centuries. Ed. by E.S.Pearson, London (Charles Griffin&Co.).

Pedersen, Olaf
1963 "The 'Philomaths' of 18th Century England." Centaurus 8(1963), 238-262.

Phillips, Derek L.
1975 "Wittgenstein und die Soziologie der Mathematik." In: Stehr, N. and R.König (Eds.), Wissenschaftssoziologie. (Köl.Z.Soziologie, Sonderheft 18/1975) Opladen (Westdeutscher Verlag), 62-78.

Pyenson, Lewis
1977 "'Who the Guys were':Prosopography in the History of Science." Hist.Sci. 15(1977), 155-188.

Pyenson, Lewis

1979 "Mathematics, Education, and the Göttingen Approach to Physical Reality, 1890-1914." Europa 2(1979), 91-127.

Pyenson, Lewis and Douglas Skopp

1977 "Educating Physicists in Germany circa 1900." Soc.Stud.Sci. 7(1977), 329-366.

Reid, Constance

1970 Hilbert. Berlin (Springer).

1976 Courant in Göttingen and New York. New York (Springer).

Richards, Joan L.

1977 "The Evolution of Empiricism: Hermann von Helmholtz and the Foundations of Geometry." Brit.J.Phil.Sci. 28(1977), 235-253.

1979 "The Reception of a Mathematical Theory: Non-Euclidean Geometry in England 1868-83." In: Barnes, B. and S.Shapin (Eds.), Natural Order: Historical Studies of Scientific Culture. London (Sage), 143-166.

1980 "The Art and the Science of British Algebra: A Study in the Perception of Mathematical Truth." HM 7(1980), 343-365.

Ross, Richard P.

1975 "The Social and Economic Causes of the Revolution in the Mathematical Sciences in Mid-Seventeenth-Century England." J.Brit.Stud. 15(1975), 46-66.

Schimank, Hans

1941 "Die Kunst-Rechnungs-liebende Societät als Gründung deutscher Schreib- und Rechenmeister." Mitt.Math. Ges.Hamburg 8,p.III (1941), 22-54.

Schimmack, Rudolf

1911 Die Entwicklung der mathematischen Unterrichtsreform in Deutschland. Leipzig (Teubner).

Schlote, Karl-Heinz

1980 "Einige Aspekte der verstärkten Hinwendung der Mathematik zur Praxis in der Zeit um 1900." NTM 17(1)(1980), 15-22.

Schneider, Ivo

1969 "Verbreitung und Bedeutung der gedruckten deutschen Rechenbücher des 15. und 16. Jahrhunderts." In: Schmauderer,E.(ed.), Buch und Wissenschaft. Düsseldorf (VDI-Verlag), 289-314.

1970a "Die mathematischen Praktiker im See-, Vermessungs- und Wehrwesen vom 15. bis zum 19. Jahrhundert." Technikgesch. 37(1970), 210-242.

1970b "Der Proportionalzirkel. Ein universelles Analogrecheninstrument der Vergangenheit." Abhandl.Ber. Deut.Mus. 38(1970), Heft 2, 1-71.

1977 "Der Einfluß der Praxis auf die Entwicklung der Mathematik vom 17. bis zum 19. Jahrhundert." ZDM 9(1977), 195-205.

Schneider, Ivo and Karin Reich

1978 "Die wirtschaftliche Entwicklung des Mittelalters im Spiegel der arithmetischen Aufgabensammlungen und ihrer Nachfolger, der Rechenbücher des 15. und 16. Jahrhunderts." Aus dem Antiquariat 1978, 6, A217-A229.

Schubring, Gert

1980 "Bedingungen der Professionalisierung von Wissenschaft. Eine vergleichende Übersicht zu Frankreich und Preußen." Lendemains 19(1980), 125-135.

1981 "On Education as a Mediating Element Between Development and Application: The Plans for the Berlin Polytechnical Institute (1817-1850)." In: Jahnke/Otte (1981), 269-284.

Segal, Sanford

1980 "Helmut Hasse in 1934." HM 7(1980), 46-56.

Shapin, Stephen and Arnold Thackray

1974 "Prosopography as a Research Tool in the History of Science: The British Scientific Community 1700-1900." Hist.Sci. 12(1974), 1-28.

Shinn, Terry

1979 "The French Faculty System, 1808-1914: Institutional Change and Research Potential in Mathematics and the Physical Sciences." Hist.Stud.Phys.Sci. 10(1979), 271-332.

Stäckel, Paul

1915 Die mathematische Ausbildung der Architekten, Chemiker und Ingenieure an den deutschen Technischen Hochschulen. Leipzig (Teubner).

Steiner, Hans-Georg

1976 "Zur Entwicklungsgeschichte der Mathematiklehrerausbildung im 19. Jahrhundert bis zum ersten Weltkrieg." In: Zur Situation der Didaktik der Mathematik im Studium der Mathematiklehrer für die Sekundarstufe II. (IDM Mat.5)Bielefeld, 10-39.

1978 "Zur Geschichte der Lehrplanentwicklung für den Mathematikunterricht der Sekundarstufe." Math.Phys.Semesterber. 25(1978), 172-193.

Stephanitz, Dieter von

1970 Exakte Wissenschaft und Recht. Berlin(De Gruyter).

Stern, Nancy

1978 "Age and Achievement in Mathematics: A Case Study in the Sociology of Science." Soc.Stud.Sci. 8(1978), 127-140.

Stichweh, Rudolf

1977 Ausdifferenzierung der Wissenschaft - Eine Analyse am deutschen Beispiel. (Report Wissenschaftsforschung 8) Bielefeld (Universität Bielefeld).

Struik, Dirk J.

1942 "On the Sociology of Mathematics." Sci.Soc. 6(1942), 58-70.

1948 A Concise History of Mathematics. New York (Dover); German: Abriß der Geschichte der Mathematik. 5th ed. with appendix: Pogrebysski,I., "Das zwanzigste Jahrhundert." Berlin (VEB Deutscher Verlag der Wissenschaften) 1972; Italian: Breve storia della matematica. Appendix on 19th and 20th century by U. Bottazzini, Bologna (Il Mulino) 1981.

Struik, Dirk J.

1980 "The Historiography of Mathematics from Proclos to Cantor." _NTM_ 17(2)(1980), 1-22.

Stuloff, Nikolai N.

1966 "Die mathematischen Methoden im 19.Jahrhundert und ihre Wechselbeziehungen zu einigen Fragen der Physik." _Technikgesch._ 33(1966), 52-71.

Sturm, Rudolf

1911 "Geschichte der mathematischen Professuren im ersten Jahrhundert der Universität Breslau (1811-1911)." _Jahresber. Deut.Math.Ver._ 20(1911), 314-321.

Swetz, Frank

1974 _Mathematics Education in China: Its Growth and Development._ Cambridge, Mass. (MIT Press).

Tarwater, J.D. (Ed.)

1977 _The Bicentennial Tribute to American Mathematics._ Washington.

Tarwater, J.D. _et al._ (Eds.)

1976 _Men and Institutions in American Mathematics._ Lubbock (Texas Tech.Press).

Taton, René

1953 "The French Revolution and the Progress of Science." _Centaurus_ 3(1953), 73-89.

Taylor, E.G.R.

1954 _The Mathematical Practitioners of Tudor and Stuart England._ Cambridge (Cambridge Univ.Press).

1966 _The Mathematical Practitioners of Hanoverian England._ Cambridge (Cambridge Univ.Press).

Thuillier, P.

1977 "Dieu, Cantor et l'infini." _La Recherche_ 84(1977), 1110-1116.

Timerding, Heinrich E.

1914 _Die Verbreitung mathematischen Wissens und mathematischer Auffassung._ (_Die Kultur der Gegenwart_, 3.Teil,1.Abt.,2.Lieferung, ed. by F.Klein _et al._) Leipzig (Teubner).

Tobies, Renate

1979 "Zur internationalen wissenschaftsorganisatorischen Tätigkeit von Felix Klein (1849-1925) auf dem Gebiet des Mathematikunterrichts." NTM 16(1)(1979), 12-29.

Tropp, Henry S.

1976 "The Origins and History of the Fields Medal." HM 3(1976), 167-181.

Turner, A.J.

1973 "Mathematical Instruments and the Education of Gentlemen." Ann.Sci. 30(1973),51-88.

Turner, G. L'E. (Ed.)

1976 The Patronage of Science in the Nineteenth Century. Leyden (Noordhoff).

Turner, R. Steven

1971 "The Growth of Professorial Research in Prussia, 1818 to 1848 - Causes and Context." Hist.Stud. Phys.Sci. 3(1971), 137-182.

1974 "University Reformers and Professorial Scholarship in Germany 1760 - 1806." In: Stone, L.(Ed.), The University in Society. Vol.II, Princeton, 495-531.

1981 "The Prussian Professoriate and the Research Imperative, 1790 - 1840." In: Jahnke/Otte (1980), 109-121.

Uebele, Hellfried

1972 Mathematiker und Physiker aus der ersten Zeit der Münchener Universität. J.L.Späth, Th. Siber und ihre Fachkollegen. Diss. Univ. München.

Van Egmond, Warren

1976 The Commercial Revolution and the Beginnings of Western Mathematics in Renaissance Florence, 1300 - 1500. Diss. Indiana Univ.

Waddington, Conrad Hal

1973 Operations Research in World War II. London.

Wall, Byron E.

1978 "F.Y.Edgeworth's Mathematical Ethics. Greatest Happiness with the Calculus of Variations." Math.Intellig. 1(1978), 177-181.

Wallis, Peter J.
1972 "British Philomaths - Mid-Eighteenth Century and Earlier." Centaurus 17(1972/3), 301-314.

Wallis, Peter J. and Ruth Wallis
1980 "Female Philomaths." HM 7(1980), 57-64.

Wilder, Raymond L.
1968 Evolution of Mathematical Concepts. London (Wiley).
1974 "Hereditary Stress as a Cultural Force in Mathematics." HM 1(1974), 29-46.

Wußing, Hans
1958 "Die Ecole Polytechnique - eine Errungenschaft der Französischen Revolution." Pädagogik 13(1958), 646-662.
1961 "Zum Charakter der europäischen Mathematik in der Periode der Herausbildung frühkapitalistischer Verhältnisse." Mathematik, Physik, Astronomie in der Schule 8(1961), 519-532, 585-593.
1979a "Zur gesellschaftlichen Stellung der Mathematik und Naturwissenschaften in der Industriellen Revolution." In: Lärmer, K. (Ed.), Studien zur Geschichte der Produktivkräfte. Berlin (Akademie-Verlag), 55-68.
1979b Vorlesungen zur Geschichte der Mathematik. Berlin (VEB Deutscher Verlag der Wissenschaften).

Wußing, Hans and Wolfgang Arnold (Eds.)
1975 Biographien bedeutender Mathematiker. Berlin (Volk und Wissen).

Wuthnow, Robert
1979 "The Emergence of Science and World Systems Theory." Theory Soc. 8(1979), 215-243.

Zetterberg, J.Peter
1977 'Mathematicall Magick' in England: 1550 - 1650. Diss. Univ. of Wisconsin 1977.

ADDRESSES OF AUTHORS

David Bloor
Science Studies Unit
University of Edinburgh
34 Buccleuch Place
Edinburgh EH8 9JT, Scotland

Henk Bos
Math. Instituut
Rijksuniversiteit Utrecht
Budapestlaan 6
Utrecht, The Netherlands

Umberto Bottazzini
Via Plutarco 12
20145 Milano, Italy

Philip C. Enros
Institute for History and
Philos. of Science
University of Toronto
Toronto M5S 1A1, Canada

Horst-Eckart Gross
Ludwig-Lepper-Str. 26
D-4800 Bielefeld, FR Germany

Thomas Hawkins
Dept. of Mathematics
Boston University
264 Bay State Road
Boston MA 02215, USA

Luke Hodgkin
Dept. of Mathematics
Kings College
Strand, London WC 2, England

Hans Niels Jahnke
Michael Otte
Institut für Didaktik der
Mathematik
Universität Bielefeld
Postfach 8640
D-4800 Bielefeld, FR Germany

Albert C. Lewis
Humanities Research Center
University of Texas
Box 7219
Austin, Texas 78712, USA

Herbert Mehrtens
Institut für Philosophie
Technische Universität Berlin
Ernst-Reuter-Platz 7
D-1000 Berlin 10

Leo Rogers
Digby Stuart College
Roehampton Institute of
Higher Education
London SW15 5PH, England

Ivo Schneider
Institut für Geschichte der
Naturwissenschaften der
Universität München
Deutsches Museum
Postfach
D-8000 München 26, FR Germany

Gert Schubring
Institut für Didaktik der Mathematik
Universität Bielefeld
Postfach 8640
D-4800 Bielefeld, FR Germany

Dirk J. Struik
52 Glendale Road
Belmont MA 02178, USA

Further Participants of the Workshop

O. Adelman (Paris)
O. Blumtritt (Berlin)
J.W. Dauben (New York)
V. DeVecchi (Toronto)
E. Garber (Stony Brook)
I. Grattan-Guinness (Enfield)
J. Høyrup (Roskilde)
E. Knobloch (Berlin)
H. Lindner (Berlin)
D. MacKenzie (Edinburgh)
H. Poser (Berlin)
L. Pyenson (Montreal)
E. Scholz (Bonn)
C.J. Scriba (Hamburg)